The Economics of On-Farm Conservation of Crop Diversity in Ethiopia

DEVELOPMENT ECONOMICS AND POLICY

Series edited by Franz Heidhues and Joachim von Braun

Vol. 45

PETER LANG
Frankfurt am Main · Berlin · Bern · Bruxelles · New York · Oxford · Wien

The Economics of On-Farm Conservation of Crop Diversity in Ethiopia

Incentives, Attribute Preferences and Opportunity Costs of Maintaining Local Varieties of Crops

Edilegnaw Wale Zegeye

PETER LANG
Europäischer Verlag der Wissenschaften

Bibliographic Information published by Die Deutsche Bibliothek
Die Deutsche Bibliothek lists this publication in the Deutsche Nationalbibliografie; detailed bibliographic data is available in the internet at <http://dnb.ddb.de>.

Zugl.: Bonn, Univ., Diss., 2004

Printed with the support of the
German Academic Exchange Service (DAAD).

D 98
ISSN 0948-1338
ISBN 3-631-53142-7
US-ISBN 0-8204-7378-2

www.peterlang.de

DEDICATION

This thesis is dedicated to the life of my mother, Yeshalem Tikuye, and the memory of my father, Getachew Wale

FOREWORD

The conservation of crop and animal genetic resources has received increasing importance in policy-making at local, national and global levels. Maintaining crop genetic resources in the form of crop varieties is topical worldwide as these are invaluable resources for addressing future demand emanating from unforeseen agricultural problems. These varieties may be needed to fight pests, diseases and solve other adaptability problems. They are especially relevant for Ethiopia that has a natural richness of crop diversity and has to strike a balance in using local and improved crop varieties to address its development and conservation problems.

Despite their importance, markets usually fail to create incentives for conserving crop genetic resources. As the socio-economic environment changes, certain variety attributes (such as environmental adaptability and yield stability) may lose their importance in the choice basket of farmers who are mainly concerned with yields. This, in turn, may endanger local varieties and induce farmers to adopt genetically uniform improved varieties resulting eventually in a loss of crop genetic diversity. This trend is aggravated by agricultural policies promoting the use of improved varieties in the absence of institutions that support the conservation of genetic information. Hence, policy makers have to also design institutions and incentive mechanisms to maintain the required crop diversity and mitigate the impact of development interventions.

There are various in situ and ex situ mechanisms to conserve crop genetic resources. This volume is mainly concerned with on-farm conservation for reasons dwelled therein. The study provides relevant information to help identify optimal policies for on-farm conservation taking sorghum, coffee, and wheat in Ethiopia as empirical examples.

To effectively design policies for on-farm conservation, the volume argues that an improved understanding of farmers' incentives, attribute preferences and opportunity costs is indispensable. These issues are extensively addressed (both theoretically and empirically) with a strong policy focus that is guided by the question: '*Given the socioeconomic set-up, what policy options are available to undertake on-farm conservation of crop diversity in Ethiopia*?'.

The study selects households most willing to undertake *de facto* conservation, identifies the most important variables that condition their variety attribute preferences, estimates average opportunity costs, and examines the factors affecting the opportunity costs across farm households. In linking the household level results to the national level, it estimates on-farm conservation costs, analyses the optimum level of investment for on-farm conservation, identifies the stakeholders to handle on-farm conservation and the incentives and institutions required. Over-all, the study addresses relevant and burning questions and offers policy options to implement crop diversity conservation on farmers' fields in the context of developing countries like Ethiopia.

Prof. Dr. Dr. h.c. Franz Heidhues
Center for Tropical Agriculture
University of Hohenheim
Stuttgart

Prof. Dr. Joachim von Braun
Director General
International Food Policy Research
Institute, Washington, D.C.

TABLE OF CONTENTS

LIST OF TABLES

LIST OF FIGURES

LIST OF ABBREVIATIONS

AAU	Addis Ababa University
BioECON	Biodiversity and Economics for Conservation
BNL	Binomial logit
CBD	Convention on Biological Diversity (1992)
CBDs	Coffee berry disease
CGBs	Community gene banks
CPRs	Common-pool resources
CSA	Central Statistical Authority, Ethiopia
CTA	Coffee and Tea Authority, Ethiopia
CPRs	Common-pool resources
EPA	Environmental Protection Authority, Ethiopia
FDRE	The Federal Democratic Republic of Ethiopia
FIML	Full information maximum likelihood
FVs	Farmers' variety (ies)
GEF	Global Environmental Fund
GM	Gross margin
IBCR	Institute of Biodiversity Conservation and Research, Ethiopia
IFPRI	International Food Policy Research Institute
IPGRI	International Plant Genetic Resource Institute
IPGs	Impure public goods
IVs	Improved variety (ies)
LDCs	Less developed countries
MNL	Multinomial logit
MoA	Ministry of Agriculture, Ethiopia
PAs	Peasant Associations
TRIPS	Trade Related Aspects of Intellectual Property Rights
UFVs	Users of Farmers' variety (ies)
UIVs	Users of improved variety (ies)
UPOV	International Union for the Protection of New Varieties of Plants (1978)
USAID	United States Agency for International Development

ACKNOWLEDGEMENTS

First and foremost let me praise and honor the Almighty God for the opportunity and capacity given to me to realize my aspirations.

Writing a Ph.D. thesis involves by and large driving a boat by your own. But one will not dare to maintain this perception soon after starting to write acknowledgements. What the boat carries and where it should go is guided, supported and inspired by many individuals and organizations. Truly, attributing the remaining errors and omissions to myself, there are many of them who have contributed directly or indirectly in the making of this thesis. Mentioning only few of them does not imply any thing more than my failure to list them all for space and memory reasons.

First of all, I extend my heart-felt gratitude to my principal supervisor, Professor Karin Holm-Müller. I am grateful not only for her stimulating and inspiring comments and suggestions but also for her kindness and availability to my enquiries without any formal process. I thank her for taking-over the responsibility from Professor Joachim von Braun with whom I started my study. I have benefited from his sharp and constructive guidance during the proposal writing. I would like to gratefully thank Professor Manfred Zeller, my second supervisor, for his constructive criticisms and comments that have been helpful to improve the content and style of presentation. I also wish to thank Professor Thomas Heckelei.

Dr. Detlef Virchow, my tutor at ZEF, was instrumental in shaping the thesis from the inception to what it is now. Identifying the topic and developing it would not have been possible had it not been for his tireless inputs. Many thanks indeed, Detlef, for your interrogative but still funny supervision. At this juncture, I wish to thank Dr. John Mburu, senior researcher at ZEF, for cooperation on a couple of conference papers.

May I gratefully acknowledge the financial support of DAAD and BMZ. The leave of absence given by Alemaya University, my previous employer, is appreciated. Moreover, organisations including CSA, IBCR, CTA, AAU, and EPA have cooperated in supplying the necessary data. The Hararghe, Jimma and Dire Dawa Bureaus of Agriculture were full of help in facilitating the household survey.

Above all, I wish to thank the interviewed farmers who were challenging at times as they did not see any livelihood improvement from repeated interviews. Had it not been for the honesty of the farmers and commitment and professional integrity of the enumerators who did the interview and data entry, the data generation would not have been possible.

A special thank goes to my good friend in Wageningen, The Netherlands, Asmare Yalew, who was willing to spare his energy and time to read most of the chapters of the first draft. I also wish to gratefully thank Dr. Assefa Admassie and Dr. Alemu Mekonnen, AAU, who provided valuable comments on the proposal. I thank Dr. Melinda Smale of IFPRI / IPGRI for her invaluable comments on the empirical chapters.

On top of these, the thesis has benefited from the comments of the participants of the National workshop on Technological Progress in Ethiopian Agriculture, First International Conference on the Ethiopian Economy, Third BioECON workshop, the IAAE 25th conference, and Tropentag conference (2003) where papers based on the thesis work were presented.

I wish to thank Ethiopian and other friends in Bonn and Köln whose company has been all the time full of joy and fun. It was a great pleasure to enjoy the company of Ethiopian compatriots at ZEF, the so called '*The Ethiopian mafia*', and the refreshing communication with all my colleagues and friends at Alemaya University. All ZEF staff and doctoral students have made the working environment collegial. Kerstin Zander, ZEF-B doctoral student, was kind enough to translate the English summary to Deutsch. I thank Dr. Manfred Denich for proof-reading the Deutsch summary and invaluable comments on the coffee questionnaire. Special thanks go to my room-mates, Nurullo and Isaak, whom I shared a lot of thoughts with.

All the staff members of the Doctoral Studies Program (especially Dr. Günther Manske and Hanna Peters) were readily available to help on administrative issues all the way through.

My brothers and sister Belainew, Dehininet, and Fassika deserve special thanks. Mulualem Getie was always at my disposal to keep me in touch with my mother. Messages from my brothers Dr. Taddesse Wale and Yisahak Abraham were refreshing and full of spirit. Last but not least, I am highly indebted to my mother, Yeshalem Tikuye, who has supported me all along my way in all respects. Thank you so much for having all the confidence and pleasure upon me, your first son. Your prayers have been a source of hope and integrity for the family and I always pray for your life.

Edilegnaw Wale Zegeye, Bonn, December 2003

1. Introduction

1.1. General background

Diversity is a reality in human existence. One could speak of diversity with respect to peoples' way of life: attitudes, culture, educational background, tastes, and preferences, for example. Diversity is also apparent in academic disciplines, political and ideological systems, language, income sources, ecology, and biological resources. In almost all the cases, the concept of diversity involves numbers and differences among the elements of the subjects of interest. Diversity in itself is resource as it adds intrinsic welfare to society's heterogeneous tastes and preferences. Diversity of all sorts offers human kind the natural freedom to choose and adjust to circumstantial endowments. Since desires, preferences and values in a given society are diverse, diversity enables mankind to match the options available with the prevailing conditions determining preferences. Diversity is, therefore, a concern for mankind. This thesis is mainly concerned with crop diversity. Definitions of different biodiversity concepts and terms frequently used are contained in Appendix 1.

Water, soil, air and genetic resources are the four major natural resources that society depends upon for food, clothing, shelter, medicament, entertainment, and even identity. Biological diversity, among other things, has ecological, genetic, social, economic, scientific, educational, cultural, recreational and aesthetic values. To capture all these benefits, maintaining biological resources is in the interest of every country.

Some countries with a high amount of unique crop diversity belong to the group of poorest countries in the world (von Braun and Virchow, 1996). Moreover, the nations with the highest biological diversity are frequently those with the least economic means to implement conservation programs (McNeely, 1988). Ethiopia is among those countries that are economically poor but still rich in biological diversity. The country has immense wealth of wild plants, animals and microbial genetic resources (FDRE, 1 998; FAO, 1998; Virchow, 1999a). It is one of the ten Vavilovian world centers of diversity around the globe identified about 80 years ago by the Russian Botanist N.I.Vavilov (Hawkes, *et.al*, 2000). It is a center of origin and as well as a center of diversity for many crops including sorghum, teff (*Eragrostis abyssinica*), coffee (*arabica*), and ensete (*ensete ventricosum*).

As a result, the country is mostly described as a land of crop diversity (Harlan, 1969).

The diverse agro-ecological geography of the country and its cultural diversity create several ways of isolating mechanisms conducive to rapid differentiation resulting in high levels of biological diversity. Even though the country has such an enormous genetic resource stock, agriculture is still unable to generate sufficient food and fibre for the population. The available stock is dwindling mainly due to changes in incentive structures in agricultural production.

In this study, sorghum, coffee and to a lesser extent wheat are considered for the empirical analysis. The relative importance of the crops in the national economy, their endemic nature, and their contrasting features are major reasons for examining the diversity of these crops. The comparison is sought with the expectation of contrasting decision making behavior of cash crop and food crop farmers. This difference, in turn, is expected to impact on the level of *de facto* conservation of the diversity of the respective crops. The growth pattern of the crops (annual versus perennial) and the market for which they are produced (only domestic versus domestic and international) could be other areas of interest for comparison.

The subsequent part of this introductory chapter is structured as follows. In the first three sub-sections, the rationales of a concern on crop diversity, the problems to be investigated, and the objectives of the thesis are outlined. Subsequently, the expected outputs, the scope of the thesis and its organization are presented.

1.2. Why a concern on crop diversity?

Improved varieties (IVs) are the crucial inputs for the development of agriculture. Crop diversity is the basic raw material used by plant breeders to develop these technologies. Breeding is a time consuming process. By the time breeders come up with a certain variety addressing a certain production problem (for example, disease), another bunch of production problems arise. As new pest and disease problems emerge and as climate changes, there is a need to evaluate useful gene sources for breeding programs to meet these newly emerging conditions (Vaughan and Chang, 1992). Breeding programs are mainly dependent on the available genetic information in order to adequately deal with potential problems facing agriculture. By conserving these resources, society can enjoy their option value as raw materials for breeding to deal with any potential agricultural problems that could arise in the foreseeable future.

Conservation of crop genetic resources, therefore, safeguards agriculture from potential losses through its future use value. To cite a practical and remarkable instance, the Barley Yellow Dwarf Virus resistant barley strain, that has been collected from the Ethiopian collection and introduced into the genetic material of the California barley in the 1960s, is saving California State from annual expenditure of 160 million US dollars for pest control (Witt, 1985). Apart from the less vivid environmental value of avoiding use of pesticides, the huge cost avoided is the outcome of maintaining local varieties of barley by Ethiopian farmers.

Crop diversity, in simple terms, protects farmers, nations and the world at large against potential hazards such as disease outbreaks, drought, and pests as it can be seen in the Californian barley instance. This protection is the result of diversified variety portfolio having different genetic make-up and potentials to withstand unforeseen agricultural problems. Because of the natural portfolio, crop diversity in agriculture provides numerous production and consumption benefits. These include better agronomic performance in different environments, insurance value (against pests, pathogens and vagaries of climate), and broad chance of addressing multiple concerns. Narrow genetic stock implies higher susceptibility for any potential danger especially in countries like Ethiopia where agriculture, the main stay of the economy, depends primarily on nature. For instance, if all farm households in a given area plant a single local variety of sorghum and if a certain disease occurs in that area which the variety can not resist, then the production loss or protection cost will have drastic socio-economic and agro-ecological consequences. To avoid such scenarios, broadening the biological resource base is a key policy issue.

One of the use value of crop diversity for Ethiopian smallholders is to enable them to survive in marginal areas. The indigenous cultivars of the various crop species found in Ethiopia are believed to possess various useful attributes suited to low-input agriculture (Worede, 1997). Their use value in agricultural research is their service as inputs for current breeding activities. Various studies have estimated that more than 50% of the increase in crop production has been due to the improvement in crop cultivars (Hawkes *et.al,* 2000).

To sum up, apart from their ecological functions (like *resilience*), crop genetic resources play key roles in the economy. They provide insurance value by reducing or spreading risks. They help to diversify products to meet diverse needs and stabilize incomes. They serve as inputs for breeding and pharmaceutical industries. They serve as sources of traditional medicine. Additionally, they reduce farmers' dependence on external inputs. Their conservation and sustainable use is, therefore, vital for the sustainable development of every country's agriculture.

1.3. The problems under investigation

According to a recent study (Worede *et.al*, 2000), Ethiopia's genetic resources are subject to serious erosion and irreversible losses. Natural resources *i.e.* land, water, forests and other forms of biodiversity have now deteriorated in the country (FDRE, 1997; 1998). Despite the resource degradation that has been occurring over the course of time, the country has still a diverse wealth of plant genetic resources (FDRE, 1996). It has also well established crop improvement program to make use of its crop diversity.

With the understanding of its complexity and multidimensionality, both ecologists and economists have attempted to explain as to why biodiversity loss occurs (Perrings, *et.al*, 1995; Wood, *et.al*, 2000). The major conclusion of most of these studies is that the prevailing incentives encourage resource users to ignore the effects of their actions. Much of the reasons are attributed to the changes in agricultural and land use practices driven by changes in the utilities of crops and preferences for different attributes. Changes in the incentive structure, markets and technological progress are favoring or disfavoring crops or varieties from the available choice set.

There are many potential tradeoffs involved between development interventions and conservation of crop diversity. As agriculture develops, it becomes more commercialized driven by markets that exist only for the crop not the genetic information contained therein. Changes in market access, technological progress, relative prices, income levels, commercialization, agricultural extension, and cash crop farming are all development interventions that will result in changes in farmers' endowments and concerns. Such actions may increase the opportunity cost of growing farmers' varieties (FVs) and endanger crop diversity in the long-run if the *status quo* is left as it is. Hawkes and his co-workers (2000) have referred to this as the *conservation / development paradox*.

Considering development as the rebalancing of the portfolio of societal assets (Solow, 1974), Swanson (1995) argues that diversity decline is the outcome of society's attempt to benefit from increasing returns using methods of mass production that involve conversion of natural capital to uniform bundle of assets. Since the advent of agriculture, societies have chosen the portfolio of living assets on which they will rely, rather than using what nature had allocated to that territory. Accordingly, it is the increasing returns from the use of methods of mass production in agriculture that drives the use of uniform varieties in agricultural production. Thus, the continued existence of most species will depend upon their capacity to find a role in the societal asset portfolio. To ensure the survival of local varieties, policy may have to change the *status quo* (by implementing different conservation programs) to realize the benefits mentioned in Section 1.2 above.

Farmers have different ways of combining their resources to achieve their livelihood strategies and each pathway has its respective consequence not only on their livelihoods but also on the natural resource base. Thus, the most important issue is which factors induce rural communities to achieve environmentally sustainable pathways of development while at the same time improving their incomes to alleviate poverty (Zeller *et.al*, 1999). Any combination of IVs and FVs that farmers take are considered here as their crop production pathways.

Mostly, conservationists seem to suggest that policies should not be biased against FVs. On the other extreme, development concerns motivate practitioners to promote efficiency and productivity in any possible way among which using IVs along with recommended technologies (such as fertilizer) is at the forefront in agriculture. The fact that the IVs are the products of FVs obliges ensuring their survival. On the other hand, IVs are the outcomes of huge research investment costs that have to be used to derive the benefits of incurring such costs. The issue at stake is deriving the benefits of IVs without sacrificing on-farm crop diversity.

Had it not been for the interest of individual farmers to maintain and use local varieties, *de facto* conservation would not have prevailed. However, policy can not entirely depend on *de facto* conservation for at least three reasons. In the first place, a farmer is not considering the size and type of local varieties other farmers are maintaining on-farm. In other words, no farmer produces diversity for its own sake. Even society does not produce crop diversity for its own sake (Hawkes *et.al*, 2000). Secondly, no farmer takes into account the effect of his/her actions on the level of crop diversity at the regional or national level. Lastly, farmers' variety choice decisions are based on observable characteristics that may not necessarily reflect all the aspects of diversity that have to be maintained. These criteria can not necessarily be expected to satisfy public conservation criteria until policy finds a way of harmonizing farmers' criteria with policy criteria. It would be sheer coincidence if the level, composition and quality of diversity corresponded to the one demanded by society (Maier and Shobayashi, 2001). The outcome of these realities calls for additional policy measures. What options are available to preserve these resources and why is the economics of on-farm conservation taken up as a research topic here?

The country has different options to conserve its genetic resources. Establishing public *ex situ* conservation sites (cold room gene banks, botanical gardens, agricultural research stations, and field gene banks run by the government), *in situ* conservation sites (farmers' *de facto* conservation, conservation on farmers' fields through external incentives, national parks, national reserves, and public conservation sites) are the routes that one can think of. Although *ex situ* conservation is still dominantly utilized, *in situ* conservation is recently being exercised in pilot projects such as the GEF on-farm conservation project in Ethiopia (Demissie and Arega, 2000).

What are the *cons* and *pros* of *ex situ* and *in situ*? *Ex situ* conservation makes it relatively easy to identify the genetic diversity conserved in the gene bank. This is possible because of full documentation for potential use by plant breeders and other scientists. Moreover, the genetic diversity maintained by these methods is directly controllable and the likelihood of losing a genetic material once it is maintained is relatively low. Moreover, access is also generally straightforward and easy.

There are also disadvantages associated with *ex situ*. Firstly, *ex situ* conservation removes genetic material from its natural environment. This halts the on-going evolutionary processes that would otherwise help to make local varieties adaptable to changing environments. Moreover, *ex situ* conservation can be capital (technology) intensive and *in situ* can be relatively inexpensive (Worede, 1997). Another criticism of *ex situ* with micro-economic rationale is that failure to utilize the gene bank at its full capacity and failure to use accessions by breeding programs means resource wastage. This has been the case for Ethiopia (See Sub-section 8.3.2). Last but not least, *ex situ* collection could also experience loss because of genetic drift (Brush, 1991).

One of the attractive features of *in situ* conservation is its conservation of both genetic material and the processes that give rise to the diversity. *In situ* also maintains evolutionary processes and farmers' indigenous knowledge that cannot be protected in gene banks (Brush, 1995; Oldfield and Alcorn, 1987; Bellon and Taylor, 1993). Moreover, the chance to conserve a lot of diversity is more open for *in situ* conservation. Potentially, far larger amounts of germplasm may be conserved on-site than is economically feasible *ex situ* (Brush 1991). The dynamic nature of the conservation, its participatory nature, and the chance it gives to link crop diversity conservation with its utilization are all the desirable features of *in situ* that seem to have influenced recent conservation paradigm. On top of these, *in situ* enables conservation of diversity at all levels (Emerton, 2000).

Moreover, some plants are not convenient for *ex situ* conservation. For instance, species like banana can not produce seeds or recalcitrant seeds like coffee can not remain viable in the cold room (Virchow, 2003). Maintaining the heterogeneous genetic structure of wild populations (like coffee) in in-vitro storage facilities is very costly and time consuming (Vaughan and Chang, 1992).

Coming to the disadvantages of *in situ*, it does not guarantee maintenance of crop diversity forever because it may collapse due to war, famine, displacement or flooding. Even though *in situ* is praised for its dynamic feature, there could be a future need to have germplasms with their current status which is not possible *in situ*. *In situ* could also be risky at times. It usually takes time to convince the local community so that *in situ* conservation can get their support. Moreover, *in situ* conservation has a political economy dimension. Above all, political will is indispensable to materialize *in situ* conservation and realize its potential benefits. If the government that has established the *in situ* scheme is replaced, the *in situ* conservation could collapse as the new government is likely to come with different set of agenda. This is due to the seemingly conflict between *in situ* conservation and development. At the policy level, the dismissal of on-farm conservation follows the idea that to encourage farmers to maintain indigenous crop varieties is to promote continued poverty and backwardness (Brush, 1992a).

Thus, neither *in situ* nor *ex situ* is free from shortcomings. Putting them as a choice is, therefore, as dangerous as comparing their costs and benefits since they are producing different services (a flow and a stock) to society (See Section 2.5 for details).

Why is on-farm conservation, the sub-set of *in situ*, a research topic for this thesis? Of the different *in situ* conservation options for crop diversity, conservation on farmers' fields has recently received a considerable attention by governments, NGOs and the international community. Among other things, on-farm conservation is emerging because it is recognized that it is impractical to conserve all potentially useful genes *ex situ* for cost and technical reasons (Hawtin and Hodgkin, 1997).

Unlike *in situ*, *ex situ* conservation (especially cold room gene bank) does not lend itself to studies on incentive design as it does not involve other stake-holders. The human element is missing in *ex situ*. Similarly, in *ex situ* conservation, there is no much degree of freedom concerning conservation options. It is only a question of technology. Questions like: 'What to conserve?' 'How much to conserve?', 'How to conserve?', and 'How long to conserve?' are all technological questions for *ex situ*. While issues of institutional design, incentives, and contracts are obligatory to be addressed in on-farm conservation, cold room gene bank conservation does not require such arrangements. For all these reasons, on-farm conservation has been taken up in this thesis. Given their complementarities, emphasis on on-farm conservation should never imply that it is the only conservation strategy to be promoted.

Farmers are managing the huge stock of crop diversity of the country. Most of the *in situ* conservation in place in Ethiopia is the result of farmers' life long practices in maintaining indigenous crop varieties through *de facto* conservation. Since farmers are the major actors in on-farm conservation, it should be planned integrating them in designing the strategies. Consequently, this thesis is mainly concerned with mechanisms of designing policy incentives building upon farmers' incentives, attribute preferences and opportunity costs.

Since on-farm conservation of crop diversity can not be undertaken in a vacuum, farmers' motivations to diversify, their variety attribute preferences, and the opportunity costs involved when planting local varieties have to be understood for informed interventions aimed at designing contextual policy incentives. Why and how are these issues relevant for on-farm conservation? Informing policy on these factors contextually will enable decision-makers to understand mechanisms of influencing farmers' variety management behavior and harmonize on-farm conservation with modern technology adoption.

This research strategy should not lead to sub-optimal policy design for it builds upon variety use behavior of the major actors. The strategy acknowledges farmers' under-valuation of genetic diversity as a local or global public good. Secondly, policy recommendations are derived from the results in such a way that the incentive design enables farmers to value the role of crop diversity to their livelihoods so that they can fully cooperate with the contract. In the presence of such an incentive framework, farmers are expected to make decisions that will lead to satisfy policy demands set to address conservation goals.

1.4. Overview of research objectives and methodology

1.4.1. Research objectives

On-farm conservation does not mean using FVs on all farmers' fields throughout the country. Given the vulnerability of millions of people to food insecurity, Ethiopia can not afford to cover all fields with FVs for the sake of conserving these resources. *In situ* conservation is not a sector-wide strategy for a nation's agriculture but one targeted to a few locations (Brush, 2000). It, therefore, requires targeting farmers who can maintain selected varieties for on-farm conservation.

Despite a lot of discourse in favor of on-farm genetic resource conservation, there is no adequate contextual research done as to how it can be harmonized with agricultural productivity objectives and made feasible complement to *ex situ*. Moreover, there are no agreed set of principles yet for on-farm conservation of crop diversity (Wood and Lenne, 1997). On top of these, little is known about the feasibility of on-farm conservation of crop diversity (Meng, 1997). In the absence of sound institutional framework and promising experience, it is not, however, unusual to find a variety of notions and recommendations (Wood and Lenne, 1997) at times conflicting. Such research gaps in Ethiopia have, among others, motivated this thesis.

The over-all objective of the thesis is, therefore, to analyze farmers' incentives, preferences and opportunity costs while managing local varieties of crops. In doing so, the thesis tries to identify the most willing farm households to undertake *de facto* conservation, the so called *pro-diversity* farm households (Cromwell and van Oosterhout, 2000) and identify the most important variables that condition their variety attribute preferences.

Specifically, the thesis addresses the following objectives:

- Examine the farm household related contextual factors affecting farmers to diversify on local varieties or to specialize with few varieties;

- Study farmers' variety attribute preferences and examine their demand for local varieties;
- Quantify the opportunity costs of growing FVs and analyze the contextual factors affecting the opportunity costs; and
- Estimate *ex ante* costs of on-farm conservation at the national level.

While the first two objectives are meant to explore the factors contributing to *de facto* conservation of crop diversity, the third is about what policy can do to improve farmers' contribution and make on-farm conservation part of their livelihood strategies. Above all, the success of the theoretical and empirical analyses done in the thesis will depend on the extent to which they have contributed to the question: '*Given the socioeconomic set-up, what policy options are available to undertake on-farm conservation of crop diversity in Ethiopia*?'

Based on the empirical results, the thesis will end up with policy recommendations to undertake on-farm conservation of crop diversity in the country. In a nutshell, the thesis is hoped to assist policy makers in the areas of selecting farmers for on-farm conservation and variety adoption, designing on-farm conservation incentives, determining the level of compensation, and research priority setting. Moreover, it will generate policy relevant information on the comparative advantages of local and improved varieties. The results will, thus, be relevant not only for on-farm conservation but also for crop variety development and diffusion.

1.4.2. Research methodology

To address the empirical objectives, data from a primary rural household survey are collected in Eastern Ethiopia (for sorghum) and in South Western Ethiopia (for coffee). Further more, the survey data collected (concerning wheat) by the Economics Department, Addis Ababa University (AAU) in collaboration with United States Agency for International Development (USAID), 1999 are used. Other secondary data are used from the Institute of Biodiversity Conservation and Research (IBCR) and Ethiopian Agricultural Research Organization (EARO).

For the first and second objectives, agricultural household models that consider on-farm crop diversity or variety attribute preference as an externality of farmers' livelihood decisions are constructed using the theory of *impure public goods, theory of joint production*, the *characteristic model,* and *random utility* theory. These models will be elaborated in more details in

Section 4.6. Farmers' demand for crop diversity, their demand for local varieties and variety attribute preferences are derived from a utility-based model. This is followed by estimation of a variety of discrete choice regression models in the respective chapters. The third objective is addressed using homogeneous treatment effect models (matching and treatment regression) and switching regression model. In all the cases, the features of the dependent variables have guided our choice of the econometric models employed. The fourth objective is addressed using the information from the opportunity cost results and the IBCR data on conservation costs and accession holdings.

1.5. Scope and limitations of the study

The inter-disciplinary nature of the subject has been a challenge and sometimes a source of frustration during the course of this study. It was a challenge and at times an opportunity to think beyond economics to get into important elements of the topic. The following points are meant to show the scope and limitations of the thesis along with a brief indication of the strategies adopted.

The number of local varieties that farmers identify does not inform about the precise levels of diversity maintained. It is not necessarily related to the diversity of landraces as recognized by geneticists since these represent two different taxonomies (Smale *et.al*, 2001). In this thesis, the diversity definition is not considering any genetic information. It is all farmers' nomenclature that has, among others, been used by Teshome *et.al* (1999) and Brush *et.al* (1992). The data used do not involve real measurements but mainly reflect farmers' variety choice criteria based on their preferences, constraints and endowments.

Failure to use and integrate genetic information into economic analysis and disregarding intra-varietal diversity is linked to the primary motivation of the thesis to explain how farmers' economic decisions result in variety diversification on-farm. To address the primary research questions, using farmers' definition of local varieties is found to serve the purpose envisaged. Using data on crop genetic diversity is hard to establish the linkage even if it were available. Moreover, such data would not have utility as they are not the outcome of farmers' conscious variety management decisions.

Moreover, all the analysis done in this thesis and all the models used are static in nature meant to capture differences across farmers. The data are concerning the year 2000/2001 cropping season. Most of the variables on which data are collected, however, pertain to farmers' decision making behavior over a long period of their farming experience (e.g. number of local varieties maintained and their experience in using IVs).

1.6. Organization of the thesis

The following chapter outlines the crop diversity conservation schemes under way in Ethiopia using historical assessment of previous literature and official documents. It presents *ex situ* and *in situ* as non-substitute conservation strategies.

Chapter 3 describes the survey data employed in the empirical chapters of the thesis. Apart from describing farmers' variety preferences and land allocation decisions, the chapter also locates the role of this exercise in model building.

Chapter 4 deals with the issues involved in facing rural development problems and crop diversity conservation in the Ethiopian context. The chapter starts with the introduction of the issues, paradigms and the challenges facing policy in the attempt to conserve crop diversity in less developed countries (LDCs). This is followed by the farm and national level potential sources of crop diversity erosion. The inputs of economics in the policy discussion on conservation of crop diversity are then subsequently taken up. This part describes the economic theories relevant for the empirical chapters to come.

Chapter 5 empirically looks at contextual farm household characteristics and their role in promoting *poly-variety*. The chapter also dwells on those factors that detach the link between *de facto* conservation and farmers' survival strategies. To address its objectives, it considers sorghum diversity as *impure public good* and the outcome of *joint production*. Having done so, it develops a utility-based household model that generates crop diversity as a positive externality of farmers' survival strategies. To empirically examine the theoretical relationships set-out and test the hypotheses presented, a Poisson regression model is derived and estimated.

Chapter 6 is parallel to Chapter 5 for coffee. Its purpose is to examine farmers' variety attribute preferences, their demand for local varieties and their *in-built incentives* to diversify. Unlike Chapter 5 that uses only count diversity, this chapter uses not only count but also other proxies for coffee diversity and dwells more on the analysis of variety attribute preferences[1]. Multinomial logit and OLS (after logit transformation) models are then estimated to examine farmers' preferences for coffee variety attributes and the proportion of coffee land that they allocate to local coffee trees, respectively.

The opportunity costs of maintaining sorghum and wheat diversity on-farm is the subject of Chapter 7. Following a gross margin (GM) computation and applying homogeneous treatment effect models, this chapter generates information on the levels of compensation required (average opportunity costs). The contextual factors affecting the size of the opportunity cost are then analyzed using a switching regression model.

Drawing on previous empirical chapters, Chapter 8 considers issues beyond the boundary of farmers. This chapter starts with the factors that will determine the 'optimum' level of on-farm conservation investment. It then estimates the *ex ante* national costs of on-farm conservation based on the opportunity cost results. It also deals with institutional issues regarding the stake-holders in the incentive design and the potentials of informal institutions for crop diversity conservation. In dealing with the incentive design, the chapter employs the theory of incentives presented in Sub-section 4.6.5. Lastly, Chapter 9 closes the thesis and presents the conclusions, policy recommendations and areas of future research.

[1]The difference is mainly due to imprecision of count coffee diversity to serve the purpose and the more detailed objectives of Chapter 6 to analyse farmers' variety attribute preferences.

2. Overview of variety use and conservation in Ethiopia

2.1. Introduction

The benefits that Ethiopia may derive from its crop diversity endowments depend on how this '*rich but poor*' nation is able to address the challenges of poverty without further degrading its natural resources. It is, thus, both a challenge and an opportunity for Ethiopia to design conservation policies that enable its agriculture-based economy to make the best use of its crop diversity.

The purpose of this chapter is to give an overview of crop diversity and their conservation in Ethiopia with special reference to sorghum and coffee. This will be done through the partial review of the available literature. The remaining part of this chapter will seek to answer the following questions. How does the relative importance of use of IVs and FVs look like historically in the country? How does the diversity of crops look like? What has the country been doing regarding conservation and sustainable use of its crop diversity?

To these ends, the chapter begins with some historical account of variety use in Ethiopian agriculture. This is followed by a narrative assessment of the role of coffee and sorghum in Ethiopian agriculture. The diversity of coffee and sorghum is the subject of Section 2.3. Examining the conservation options and constraints based on historical account and current developments is the purpose of Section 2.4. Section 2.5 justifies the complementarities of *ex situ* and *in situ* to be followed by a summary in the last section.

2.2. Crop variety use in Ethiopian agriculture

Agricultural research and extension are the key actors in agricultural technology development and transfer process in Ethiopia. The history of agricultural research in Ethiopia dates back no further than early 1950's (Debela, 1988). Both agricultural extension and research are publicly funded and their performance and scope of operation has been politically motivated. Due to political in-stability, their strategies lack continuity and there have been at least as many changes in strategies as there are changes in government. This will mainly affect *de facto* conservation by farmers due to its direct link with the agricultural development strategy followed.

Ethiopia is a country still heavily dependent on FVs for its national food production (FAO, 1996; quoted in Virchow, 1999a). Over 90% of the agricultural land is characterized by FVs use (Kidane and Abler, 1994) and less than 10% of peasant farms use IVs. According to the data from Central Statistical Authority (CSA), considering the years 1990 to 1991, IVs were used only on 4 percent of the area allocated to cereals. The local varieties of many crops are still in the hands of millions of small-holder farmers for their unique role in addressing farmers' concerns and constraints. However, the performance of agriculture thus far calls for transformation through technological change.

With the hope of addressing its severe rural poverty problems, the present Ethiopian government adopted the so called Agricultural Development Led Industrialization (ADLI) in the mid 90's. Agricultural extension package is the major component of this strategy. The agricultural extension package program is supply-driven in nature in that its primary focus is on high yielding crop varieties and high potential areas with little regard to the needs and priorities of chronically food insecure areas and resource poor farmers (CRDA, 2002). Tying of credit and fertilizer supply to adoption of the so called 'extension package' has been the practice of the local development agents and extension workers for the last decade. Accordingly, agricultural extension, input supply and credit services are mainly geared towards the so called '*high potential areas*'. The conventional emphasis in research and development strategy - investing in high potential areas - is still the dominant strategy of technology development and transfer in the country. If adoption of IVs is resulting in crop diversity loss, the loss must have been felt more intensively in these localities.

The strategy of the package is assuming that IVs are profitable for all farmers operating in the so called '*high potential areas*'. But this is not necessarily the case as the comparative advantage of the IVs and FVs varies across farmers and localities (Edilegnaw, 2003). While some farmers and localities can make the best use of IVs, still others can do better with the FVs at their disposal. Similarly, Worede and others (2000) have reported the case that the local durum wheat varieties out-performed the IVs in the central high lands of Ethiopia. This can partly explain the rationale behind farmers' reluctance to adopt improved technology packages despite the huge effort in the popularization and transfer of technologies. The underlying reasons are mainly documented in the partial adoption literature (Feder *et.al*, 1985; Feder and Umali, 1993).

Thus, IVs are not a *panacea* for all farmers. Complete dependence of agriculture on either variety set is not an option. Policy has, therefore, to identify areas and farmers where the respective varieties can work better. On top of this, not just varieties but the supply of other supplementary services - credit, input supply and marketing opportunities - is indispensable.

Besides, investments in high potential areas and less-favored areas should not be seen as mutually exclusive. The government has to invest in both areas based on the prevailing conditions affecting comparative advantages. The difference is on the type of investments where less-favored areas may call for promoting the productive capacity of FVs and high-potential areas for investments in more capital intensive technologies. Otherwise, placing capital intensive technologies in less-favored areas (Hazell, 2002) for the sake of short-term productivity gains may even lead to more diversity opportunity cost because the less-favored areas are relatively rich in crop genetic diversity.

2.3. Coffee and sorghum in Ethiopian agriculture and their diversity

Coffee and sorghum are two very important crops for the livelihoods of millions of farmers in Ethiopia. Not only are the crops important to the Ethiopian agricultural economy but also is their diversity of a particular interest internationally. In this section, these two points will be elaborated.

2.3.1. Coffee in Ethiopian agriculture and its diversity

Ethiopia is the oldest coffee exporter in the world. According to the Coffee and Tea Authority (CTA), it is the third largest exporter in Africa (next to Ivory Coast and Uganda) though external invasions and internal conflicts have at times had a negative impact on the country's coffee export history (CTA, 1999). Coffee contributes in the Ethiopian economy about 10% of the GDP, 12% of the agricultural output, 70% of the country's foreign exchange earning, 10% of the government revenue, and 25 % of the livelihood of Ethiopian population (CTA, 1999).

Even though it is hard to prove, *Coffea Arabica* is said to originate in Ethiopia and of course its genetic diversity in the country is one of the highest. Ethiopian coffee contains a great genetic variation to serve as the best source of germplasm (Sylvain, 1958). Many ethno-botanical studies assert that coffee got its name from a region in Ethiopia called 'Keffa' and the trees were called '*Kafa*'.

Many studies have shown that Ethiopia's coffee diversity plays an important role not only for the country but also for the world at large (Sylvain, 1958). Ethiopia holds 6% of the world's coffee *ex situ* conservation holdings (FAO, 1998 and Hawkes *et.al*, 2000). Ethiopian coffee genetic materials are represented in world collections with numerous samples exchanged among gene-banks and breeders (Worede, 1988) and the country is the only region where *coffea arabica* is found as wild forest species (Worede, 1988; Berthaud and Charrier, 1988).

The diversity of the Ethiopian coffee genetic stock has enabled it to withstand uprising production problems. As a result, coffee production has not been threatened by leaf rust (Dubale and Teketay, 1999). The outbreak of coffee berry disease (CBDs) in 1971 did not abandon the production of coffee in the country (Tewolde, 1990) in contrast to the wiping out of plantations in Sri Lanka (in 1869) and Sumatra (in 1876) as a result of outbreak of coffee leaf rust (Dubale and Teketay, 1999). However, the genetic resources of coffee and the associated flora and fauna are disappearing rapidly as a result of deforestation of the ecosystems (Dubale and Teketay, 1999).

Despite its immense role in the economy, most conservation efforts for *coffea arabica* have not proceeded further than a proposal stage (Dubale and Teketay, 1999; Gole *et.al*, 2002). Due to its limited financial and technology capacity, the country can not conserve this genetic resource by its own (Tewolde, 1990). Still, the crop that has been given most priority in the conservation exercise in Ethiopia or worldwide through living collections is coffee (Worede, 1985). Coffee seeds can be maintained *ex situ* (either in the field gene bank or in the cold room using in-vitro facilities) and *in situ* (on farmers' fields or in the natural forest for forest coffee)[1].

Because of price collapse, coffee farmers are already starting to uproot old coffee trees and plant annual and other high value crops that can make their decisions more flexible. Replacing coffee by cht'at (Catha Edulis)[2] has been farmers' practice for long in Eastern Ethiopia (Worede and Getu, undated).

---

[1]The seed conservation of coffee germplasm is rather problematic (Dullo *et.al*, 1999) provided that its storage behavior does not generally conform to either orthodox or recalcitrant types (Ellis *et.al*, 1990).

[2]This is an important stimulant cash crop in the South Eastern part of the country. Consumers chew the leaves and ensen themselves.

Coffee price crisis affects not only the livelihood of about 15 million Ethiopians but also the coffee genetic resources as a result of the replacement of coffee trees by alternative crops (Gole, 2002). Even though there is no danger of immediate replacement of FVs by IVs, the possible replacement of coffee by other high value crops may require due attention. The emerging certified organic coffee and fair trade movements (Rice, 2001) are promising signs to reward poor farmers for their contribution and save potential coffee diversity loss.

2.3.2. Sorghum in Ethiopian agriculture and its diversity

Sorghum was named because of its height from the Latin word 'surgo' which means 'arise' (Karper and Quinby, 1947). The crop, first domesticated in Africa, is the major cereal that has a wide range of utilization not only in Ethiopia but also in many other African nations (Gebrekidan, 1979; Gebrekidan and Kebede, 1979).

In Ethiopia, sorghum is the major crop next to teff grown all over the country across high, intermediate and low altitude areas. The crop occupies 0.7 to 1.1 million hectares with 1 to 1.6 million tons of production annually. It contributes about 15-20% of total cereals production in the country. It is typically important crop in the low land areas where drought and moisture stress are recurrent. The descriptive survey data show that it is used for many purposes (food, cash, animal feed, fuel, house construction and fence) than perhaps any other crop.

Sorghum consists of over 20 different species of which some come from the East African and some from the West, Central and Southern African Region. The primary center of origin and diversity of sorghum is assumed to be the Sudan and Ethiopia (Doggett, 1991) and breeders have found useful attributes from the Ethiopian germplasm stock (Kebede, 1991). Four of the five major races of sorghum in Africa are found in Ethiopia (Stemler *et.al*, 1977). Ethiopia holds *ex situ* 4% of the world's sorghum genetic stock (FAO, 1998 and Hawkes *et.al*, 2000). The crop exists in tremendous variability throughout the areas of sorghum production in Ethiopia (Gebrekidan, 1979; de Wet and Harlan, 1971).

The threat of genetic erosion in sorghum in Africa as a result of expansion of new uniform cultivars and hybrids has not been as great as in other major cereals such as maize. Cyclical drought and crop failures have probably been the major causes of sorghum diversity loss in Africa (Gebrekidan, 1979).

2.4. Conservation of crop diversity in Ethiopia: strategies and options

Various government organizations are involved in the conservation and sustainable utilization of crop genetic resources including IBCR, EPA (Environmental protection authority), EARO, CTA, and MoA (Ministry of Agriculture). In addition, many NGOs (for instance, Ethio-Organic Seed Action – EOSA[3]) and grass-root organizations dealing with the local community are working in the areas of resource conservation and rural development. This section describes the national *in situ* and *ex situ* undertakings and potentials.

2.4.1. *Ex situ* conservation and its potentials in Ethiopia

The IBCR in conjunction with relief agencies has been trying to salvage crop diversity by taking samples from farmers' stores and markets and storing at strategic seed reserve centers in areas where drought is prevalent (Worede, 1985). We briefly consider *ex situ* conservation in Ethiopia.

2.4.1.1. Cold room seed gene bank conservation

IBCR (the then Plant Genetic Resources Center) is established in 1976 through a bilateral cooperation between the Ethiopian government and the government of the Federal Republic of Germany to rescue the country's biological resources. The initial laboratory and office building was built at a cost of over half a million Ethiopian Birr. Since its establishment, it has been responsible for the collection, characterization, conservation, documentation, evaluation and utilization of biological resources of the country. The Institute is actively involved in the global network of genetic resources conservation and utilization activities which are coordinated by FAO. It has close working relationships with international centers like ICRISAT, ICARDA, CYMMIT and CIFOR among others (Worede, 1985). The FVs collected by the Institute are being utilized in national and international crop improvement programs.

In the first ten years of its operation, the IBCR has conducted more than 76 collecting expeditions (Demissie, 1991). The capacity of the Institute is increasing in all respects. Some five years ago, it holds 55,548 accessions (Yibrah, 1998). Currently, it has about 61,000 accessions of over 101 species of crop plants including cereals, pulses, oil crops, and to some extent lesser

---

[3]This NGO can be considered as a pioneer in linking crop diversity conservation on-farm with rural livelihoods.

known but potentially useful crop plants and medicinal plants (IBCR, 2003). The Institute holds 7,778 *sorghum bicolor* accessions (Yibrah, 1998).

2.4.1.2. Conservation in the field gene bank

In Ethiopia, medicinal plants, coffee, root crops, and spices are all maintained in field gene banks at the appropriate agro-ecological sites (Demissi, 2001). For coffee, there are three important field gene banks: Chochie (Jimma), Sidamo, and Hararghe.

During the field work, Chochie coffee field gene bank was visited. The field gene bank is established in 1977 E.C. (1984 G.C.) by the then Plant Genetic Resources Center with the financial support of the GTZ on the basis of bilateral agreement. It is located 376 km from Addis Ababa in the South West part of the country. What follows is the information gathered from the gene bank manager about its status at the time of the data collection.

The field gene bank has four major blocks and has currently 119 plots. Among these plots, 50 of them have 3 trees per accession and the rest 69 plots have 6 trees per accession. In 1990, the Chochie field gene bank had 4,000 accessions (Dubale and Teketay, 1990). At the moment, the field gene bank holds 4,537 accessions containing about 25, 000 trees of coffee.

There are different problems that the gene bank is facing at the moment. Firstly, frequent drought is damaging the different accessions and shade trees. The different accessions coming from different localities have different growth pattern, tree height, maturity and so on. This has proved to create management problem as far as the application of different cultural practices (weeding, cultivation, and pruning) is concerned. Their adaptation has also been a problem.

The gene bank has tried to collect accessions from the major coffee growing areas of the country. However, some of the accessions coming from the very remote areas could not give any yield because they could not adapt to the environment and in some cases they died. Inorganic fertilizer is not used in the field gene bank. Thus, trying to conserve a large number of species at a single site to optimize costs may not be feasible in field gene bank conservation.

After giving good yield only for few years, most of the coffee trees had to be up-rooted. To alleviate this adaptation problem, every year the leaves and branches are cut and the beans are collected to improve the carrying capacity of the trees.

There is now lack of follow-up and no adequate budget to field gene banks like Chochie. The bank does not systematically monitor the costs incurred and the returns generated. As a result, data on costs and benefits could not be collected. If the gene banks in the country are well managed, they can at least cover their running costs and conserve the coffee genetic diversity (Chocie Site Manager, November 2002). The field gene bank conservation strategy has to be strengthened by identifying representative sites. Otherwise, collecting accessions from remote areas and putting them all in a single spot could create adaptation problem and result in even more loss than conservation.

2.4.2. *In situ* conservation and its potentials in Ethiopia

In situ conservation ventures are of recent innovations in Ethiopia and they are little tried (Demissie, 2001). *In situ* conservation in Ethiopia is yet at its infancy. Despite the country's huge potential in its genetic resource stock, the *in situ* conservation activities are of limited nature in both coverage and scope. This is mainly due to the limitation in resource capacity. Lack of experience and absence of well established principles are the forefront hurdles. The dominant *in situ* activities in place are yet in a pilot project form that can not necessarily proceed more than the project life. As they are not established with the understanding of farmers' motivations, expectations, endowments, and the institutional setting, they are no more important than learning from trial and error. Most of the *in situ* conservation models suggested in Ethiopia are of dubious acceptability to farming communities (Demissie, 2001). Thus, there is still a long way to go before the Ethiopian government can institutionalize *in situ* conservation and make it convincing for the poor farmers.

2.4.2.1. Conservation on farmers' fields

Farming itself is the original method of conservation linked with utilization (Engels and Wood, 1999). On-farm conservation is a process managed by farmers themselves and not one to be imposed by institutions (Wells, 1998). Farmers are custodians of genetic resources and always retain some seed stock for security unless circumstances dictate otherwise (Worede, 1988). They have always been doing so well ahead before the gene bank came into the picture. Taking farmers as custodians of crop diversity, the history of on-farm conservation can, therefore, be traced back to the history of farming.

According to some studies, the concern of losing their local varieties and breeds has always existed among farmers. There are multiple examples of farmers who maintain crop and animal genetic diversity for their own present and future use (Mugwara and Gebre Egziabher, 2002). However, one should not lose sight of the fact that farmers do not maintain local varieties for diversity sake in a way required by the public agency dealing with conservation.

Farmers' knowledge is believed to be important not only for retaining crop diversity in its natural environment but also in integrating scientific knowledge to their life styles. The Ethiopian farmers are said to be not only an adjunct to conserving and enhancing but also key actors in maintaining genetic diversity (IDRC, 1995). As we proceed, the GEF project that has served live experience with on-farm conservation in Ethiopia will be described in the relevant chapters.

2.4.2.2. Enhancing farmers' varieties

Variety enhancement, which refers to improving the productive capacity of FVs, could be done to broaden the genetic base of individual cultivars with respect to different variety attributes. Enhancing FVs is like achieving two objectives (conservation and development) with one means because this system will maintain these resources and farmers will get better utility from their varieties.

Enhancing FVs is perhaps the most feasible strategy that can be followed in most parts of the country where local varieties are in use. This strategy should attempt to study the agro-ecological and agronomic factors that determine the comparative advantages of the respective varieties (Edilegnaw, 2003). Of course, most of the breeding work in the country is basically selection of the better varieties from the available set (with respect to required attributes). This could be considered as a modified version of variety enhancement even though it does not have conservation dimension.

Genetic selection and enrichment and a link between conservation and development is recognized as the third conservation strategy complementary to *ex situ* and *in situ* conservation in the narrow sense (Alteiri and Merrick, 1987; Hardon and de Boef, 1993). Local crop development with its integration of local breeding is the basis for such conservation strategy maintaining, stimulating and enhancing the dynamics of the community management of crop diversity.

Enhancing farmers' varieties is the single most untouched strategy that can work in Ethiopia. Given the survival of most crop varieties under stress conditions, this strategy has to be further exploited to address both development and conservation problems.

2.4.2.3. Community gene banks

Two of the stated plans of the country's Biodiversity Institute are strengthening community gene banks (CGBs) and strengthening on-farm conservation across a broader range of agro-ecological conditions. Recently, it is establishing CGBs that organize the local community having both conservation and development ends.

CGBs are a simple process of community agriculture including village level facility and a garden or a field where local varieties are maintained (Feyissa, 2002). CGBs collect and buy seeds from farmers after establishing network to facilitate seed supply among the farmers. Every member farmer gets seeds of his/her choice for planting and returns the seeds at an agreed rate during harvest. CGBs have two major components *i.e.* a seed store and a germplasm repository for local crop improvement complementary to the field gene bank. They also have market component.

These seed banks are promising innovations that can help to enhance the role of farmers in the conservation exercise. They pool all the potentially valuable crop varieties so that member farmers can have access on demand. In doing so, CGBs ensure that all the FVs are in the pool year after year. Thus, they are serving as important seed reserve mechanisms.

Moreover, CGBs serve as important mechanisms to increase the number of options for using diverse crop types and to decrease vulnerability to seed shortage. In Ethiopia, they function as community-based seed supply networks for locally adapted crops and enhanced FVs (Feyissa, 2002). They are decentralized across regions. There are about 12 CGBs established in different farming systems through the financial support of the GEF.

2.5. *Ex situ* and *in situ*: why complements?

The purpose of this section is to explain the non-substitutability but complementarities of *in situ* and *ex situ* conservation strategies. In doing so, it has to be made clear that the end results are different. While *ex situ* strategies generate stock information for future decision making, *in situ* strategies supply a flow of information (Swanson and Göschl, 2000a). *In situ* and *ex situ* are, therefore, imperfect substitutes (Smale *et.al*, 2001). *In situ* is basically production (or utilization) along with conservation whereas *ex situ* conservation in the gene bank is purely meant for static conservation involving only the use of the germplasm for breeding. But still they are not mutually exclusive because they will have some common outcomes depending on factors like the rate at which *ex situ* collections are done, the communalities of the genetic resources considered for conservation *in situ* and *ex situ*, and the rate at which evolution in the field occurs.

To utilize both options efficiently and to capitalize on the dynamic features of *in situ*, one has to ask 'How fast is crop genetic variability occurring under the natural environment?' The faster the rate of variability, the more *in situ* is justified. If it takes 50 years for a genetic material to change its attribute, there is little economic and biological ground to justify *in situ* conservation of that genetic material as this duration is so long and beyond the public decision horizon of the present generation (Swanson and Göschl, 2000a). If the rate of change in genetic structure of a certain crop occurs so fast, *in situ* conservation of that crop will be sounder. This is because the difference between the stock and the flow of information will be high enough to justify the costs.

Most authors agree that *ex situ* conservation should serve as a back-up for *in situ* conservation and they should be considered as complementary (Shands, 1991; Brookfield and Stocking, 1999; Brush, 2000; Hawkes, *et.al*, 2000). Even though both *ex situ* and *in situ* conservation ventures are expensive, they are both necessary in order to approach an efficient method of resource conservation (Swanson and Göschl, 2000a). Their complementarity emanates from the fact that they are serving different conservation purposes: conservation at a point in time and conservation over time. Having this in mind, the two systems are adopted into the conservation strategy of Ethiopia (Demissie, 2001). Above all, establishing the linkage between *in situ* and *ex situ* conservation is the most crucial and challenging task ahead (Demissie and Arega, 2000). This can help avoid redundant conservation activities. Moreover, it enables policy to utilize or hold down the strengths or weaknesses of the respective conservation options.

Mostly, the question of optimum combination of *in situ* and ex situ is raised (Swanson and Göschl, 2000a). However, the question should rather be as to how to make their outcomes supportive and non-redundant. How much should a country invest in each conservation strategy basically depends on the type of genetic resources it is dealing with[4] and its resource endowment. This is mainly because each strategy needs different types of investment and each one is differently suited to different types of crop diversity.

For all the above reasons, policy makers have to recognize that *ex situ* and *in situ* are neither substitutes nor mutually exclusive in outcomes. Based on the features of the genetic resources and scarcity of the resources required to undertake the different conservation strategies, the country needs to capitalize on the complementarity of these strategies.

2.6. Summary and concluding remarks

In this chapter, an overview of crop diversity conservation in Ethiopia has been provided. From the assessment of crop variety use in Ethiopian agriculture, it has been clear that Ethiopian farmers are mainly dependent on their own local varieties.

Even though the experience in *in situ* conservation is very recent and scanty, the country is undertaking different *in situ* and *ex situ* conservation activities. Given the limited experience in *in situ* conservation, there is still a long way to go to institutionalize its implementation.

The chapter has described the strategies in place and the possible ones worth considering with their *cons* and *pros*. One compelling point to note from this attempt is that it can not be singled out and argued either for *in situ* or *ex situ* conservation without putting them contextually. It has to be seen in the light of their complementarity taking into account potentials for agricultural intensification. Moreover, on-farm conservation has to be context specific as there can be no formula that works everywhere. The next chapter moves into the data collection process and farmers' variety choice.

[4]This affects the rate at which the diversity changes under the natural conditions.

3. Data collection methodology and insights on farmers' variety choice

3.1. Introduction

What descriptive insights can be generated from the household survey regarding farmers' variety choice behavior? Taking this question as a guide, the purpose of this chapter is to describe the data generation process and examine as to what and why farmers are doing what they do regarding variety management. This is done to understand what it means for on-farm conservation and model building in the chapters to come.

To these ends, Section 3.2 describes the data generation process. Sections 3.3 and 3.4 continue to describe own survey data and the AAU / USAID data, respectively. Farmers' variety choice and land use decision for coffee and sorghum are presented in Section 3.5. This is followed by their comparative assessment of improved and local varieties in Section 3.6. Coffee and sorghum farm household decisions are briefly compared in Section 3.7. The role of understanding farmers' varietal decisions in model building is the subject of Section 3.8 followed by a brief summary at the end.

3.2. The data generation process

The empirical analyses in this thesis are based on primary and secondary data. Secondary data were collected from concerned government offices and visits were made to different gene banks and research centers. Before sampling farmers for the interview, a base-line survey was undertaken to identify sources of heterogeneity of farm households and sampled villages. Moreover, key informants were interviewed in the respective regions to identify key contextual factors prevailing - most important variety attributes, resource constraints, and agricultural problems.

For the household survey, the thesis has adopted a *multi-stage stratified sampling technique*. Selection of zones for the respective crops was purposive based on the relative importance of the crops considered. Three agro-ecologically contrasting and neighboring zones of South Eastern Ethiopia (East Hararghe, West Hararghe, and Dire Dawa Administrative council) were considered for sorghum. For coffee, Jimma administrative zone in South Western Ethiopia was considered. From each zone, the

Districts were ranked based on the relative importance of coffee and sorghum and three Districts were considered in each Zone. From each District, 2-3 Peasant Associations (PAs) were purposively selected based on their agro-ecological representativeness and importance of the crops considered. The latter data are from the respective District Bureaus of Agriculture.

In consultation with Agricultural Bureaus and development agents working with farmers, the survey sample was structured to cover a representative sample of villages and farmers with respect to a wide range of agro-ecological and economic variables (including prevalence of drought, rainfall distribution pattern, distance from markets, distance from extension and input supply services, relative importance of different income sources, and prevalence of poverty). Finding the most important sources of heterogeneity of peasant households was the most important step before the stratification. This was done in consultation with the respective development agents and basic household roaster kept in each PA. The most important sources of heterogeneity were land size, poverty status, sex of the household head, income source outside agriculture, and household size. After finding the sources of heterogeneity among farm households, they were s tratified and grouped. After stratifying the farm households in the respective groups, individual farm households were randomly sampled proportionately in each stratum to get representative-ness.

Every household in the sampling frame list cultivates sorghum or coffee. The s ample includes a reas w here c offee and s orghum p lay i mportant r ole because on-farm conservation is more sensible in areas where the production of the crop under consideration plays an important role to farmers' livelihood. Accordingly, coffee was the first most important crop for 62 percent of the sampled farm households and sorghum was the first most important crop for 76.8 percent of the sampled farm households.

The data concerning coffee were collected from 8 PAs in Mana (Harro and Kella Guda), Goma (Bulbulo, Kilole Kirkir and Yachi Urechi) and Seka Chekorsa (Gibe Boso, Hallo Sebeka and Sebeka Debiye) districts of Jimma Zone. A total of 266 farmers (on average 33 per PA) were sampled, out of which the heads of 247 households were males. The data concerning sorghum were collected from 7 PAs in East Hararghe (Belina Arba, Umerkule, and Kero Deda), West Hararghe (Chachole and Gurbo) and Dire Dawa (Asseliso and Eje Aneni) zones of South Eastern Ethiopia. A total of 198 farmers (on average 28 in each PA) were sampled of which the heads of

185 households were males. In each PA, reserve farmers were also selected in case the selected farmers are missing.

Ten enumerators were hired (five for Jimma zone and five for the others) and trained (in class and in the field). For better interaction of enumerators with farmers, it was important to select enumerators who know the local language, culture and setting. Four of the enumerators were diploma holders and the rest have completed high school.

To check the relevance of the questionnaire to the local conditions, farmers' expectations and level of understanding, the questionnaires were pre-tested on 16 farmers in each case. After the questionnaires were pre-tested, the test results were intensively discussed with the enumerators to accommodate the necessary changes in line with farmers' response to the previous version. The pre-testing was very helpful to gauge some questions and have more questions for the sake of cross-checking the validity of farmers' response.

Local development agents, who are employees of the MoA, have been assisting at every stage of the field work to get in contact with the sampled farmers. Some farmers were not able to identify the name of the FVs and most were not able to identify the different IVs. They could only say either local or improved in their own language. Once farmers identify different local and improved varieties, the development agents working in close connection with the farmers have assisted to verify variety types (local versus improved) and sort out differences and similarities of different variety names.

The household survey was conducted from July 2001 till March 2002. Interviews were conducted in places convenient for farmers (either at home or in the field). Each enumerator did an average of three interviews a day. The details of the variables on which data were collected are contained in the questionnaire presented in *appendix* V.

3.3. Description of the survey data

3.3.1. Data description for coffee growing farmers

Among the 266 coffee growing farmers, 76 percent of them are Muslim. Polygamy is very common in this study area. Some 17.4 percent of the household heads have more than one wife of which the maximum reported only by two household heads was 5.

Ninety-nine of the coffee growing farmers (37.4 percent) have participated in the coffee extension package at the time of the interview. Among the extension participants, 78 of them are still in the package and 22 have interrupted. About 98.5 percent of the coffee growing farmers are members of some local and informal institution(s). So far, the informal institutions have no influence on variety choice whatsoever.

The household head(s) and the key informants were asked to judge the relative poverty status of each household. According to the household heads, 28.2, 65 and 6.8 percent of them are poor, medium and rich, respectively. When the key informants judge the household poverty status, these figures were 30.8, 47.7 and 21.4 percent, respectively. The judgments of the farmers and the key informants coincide in 82 percent of the cases. Some of the poor, judged by the key informants, seem to categorize themselves as medium and most of the rich also put themselves as medium. Their poverty status is as diverse as their income sources to the extent that some have managed to own a car and give transport services to the local people.

Regarding their income sources, 22.3 percent of the farm households have other income sources outside agriculture. Coffee growing farmers are involved in diverse set of non-farm and off-farm economic activities including local trading, daily labor, employment in government enterprises, carpentership, pottery, weavery, traditional medicine, preparing and selling local food and drinks ('*areki*', '*tella*', '*injera*') and butchery. About 12 percent of the households have reported non-farm small scale economic activities and off-farm job opportunities as the third (next to crop and livestock production) most important sources of income. The sampled coffee growing farmers have been experiencing diverse set of survival strategies during last crop failures. Selling fixed assets, off-farm and non-farm work, migration to cities, planting early maturing or drought resistant plants like enset, looking for government aid and taking loan from relatives and friends have been the most important means for farmers to survive when crops fail.

Due to the non-flexibility of repayment dates of input supply credit from the MoA and the need to satisfy immediate household needs, most farmers are forced to sell coffee right after harvest at low prices. The data show that about 85 percent of the sampled coffee growing farmers sell coffee either during harvest or right after harvest.

3.3.2. Data description for sorghum growing farmers

Almost all (98.5 percent) of the sampled farm households for sorghum are Muslim. Some 110 of these farm households have been participating in the extension package of which only 57 are still participating. About 79.8 percent of them have been rearing livestock. Majority (91.2 percent) of the sorghum growing farmers are members of some local and informal institution(s). Like the case with coffee, so far, the informal institutions have no influence on farmers' sorghum variety choice decisions.

The results of the quick poverty status assessment by the household heads reveal that 44.7, 48.7 and 6.6 percent of sampled farmers are poor, medium and rich, respectively. When the key informants judge the household poverty status, these figures happen to be 38.9, 41.4 and 17.7 percent, respectively. The judgments of the farmers and the key informants coincide in 79 percent of the cases.

As far as their income sources are concerned, 41.2 percent of the farm households have other income sources outside agriculture. They are involved in diverse set of non-farm and off-farm economic activities including fire wood and charcoal selling, daily labor, black smith, local trading, handicrafts, selling bricks and guard. About 21 percent of the households have reported non-farm small-scale economic activities and off-farm job opportunities as the third most important sources of income. Selling fixed assets, o ff-farm and non-farm work, migration to cities, selling fuel wood and charcoal, and government aid are the most important means for farmers to survive when crops fail. Most of these activities reveal to what extent farmers' survival strategies are directly affecting the natural resource base.

About 80 percent of the sorghum farmers sell their crops during harvest or right after harvest during which prices are at their lowest margin. Like in the case of coffee, output prices are not attractive for sorghum farmers and input prices are already beyond the purchasing power of many of the farm households. As a result, most farmers are substituting food crops (like sorghum) with cash crops (like cht'at) in the Eastern part of the country.

3.4. The AAU / USAID data for wheat

With the purpose of comparing opportunity costs with sorghum, the AAU / USAID survey data concerning wheat were used. The presence of comprehensive input-output data has given a chance to compute the opportunity costs of growing local varieties of wheat.

Wheat is one of the most important crops and ranks fifth in acreage after teff, maize, sorghum and barley in Ethiopia (Engels and Haukes, 1991; Gebre-Mariam, 1991). Ethiopia is the largest wheat producer in Sub-Saharan Africa covering about 0.75 million hectares of farm land throughout the country (Gebre-Mariam, 1991). The country is also the center of diversity for durum wheat (*Triticum durum*) (Mulualem, 1999).

Taking into account the importance of the crop and to use the chance it offers for comparison of opportunity costs, the 1999 rural household survey data collected by the Economics Department, AAU in collaboration with the USAID, were used to compute and examine opportunity costs of growing local durum wheat in Chapter 7. The data involve 1681 farm households of four large regions in the country (150 in Tigrai, 466 in Amhara, 625 in Oromiya, and 440 in SNNP[1]). From among 671 plots, 634 plots which were not inter-cropped are considered in the analysis. Parts of the data that deal with 402 wheat growing farmers were extracted. Among these farmers, 352 of them were using local varieties and the remaining 50 were using improved ones.

3.5. Farmers' variety choice and land use decisions

The purpose of this section is to describe the land use and variety choice decisions of the coffee and sorghum growing farmers using the survey data.

3.5.1. Coffee variety choice and land use decisions

Coffee is basically grown for two major purposes: own consumption and addressing cash demand (clothing, tax payment, health expenditure, shelter construction, purchase of agricultural inputs, social obligations, purchasing other agricultural food crops and school fee).

Farmers very well recognize that all varieties have different potentials with respect to different attributes. Having different concerns and resource constraints, they want to know the one(s) that best fit(s) their interest. By doing so, they are making use of the potentials of the different varieties. By increasing their choice portfolio, they increase the probability of getting a set of varieties that address their concerns. Not all reasons that farmers give for

[1]Southern Nations, Nationalities, and Peoples Regional State.

maintaining local varieties are necessarily of economic nature. For instance, some farmers have stated that they keep a variety simply because their father gave it to them.

When ranking the importance of coffee variety attributes to the household, farmers have given more rank to adaptability to the environment, yield, disease resistance, yield stability, market price attraction and average yield during bad season. Even though farmers have noted that the market does not make significant price difference across varieties, they still want to have a coffee variety that can attract better price. When farmers were asked if they want to have as many varieties as possible, 92.5% of the total respondents were interested to have many varieties. Shortage of land is the prominent reason for most of those farmers having no interest with many varieties.

Those farmers who opted for many varieties would like to select the best performing variety (ies) with respect to their variety attribute preferences. The variety attribute 'food taste' is given priority not only for the purpose of household consumption but also for its capacity to attract better market. Farmers very well know that diversification helps to survive if things go wrong. Moreover, it will be simple for them to choose the best variety (ies) among a broader choice set. They also make use of the possible alternate fruit bearing to stabilize income.

When farmers were given two options (See Question 22.0, Appendix V), 42.7 percent of the farmers (mainly attracted by its convenience for management) have opted for one variety, 41.7 percent of them (mainly attracted by the chance to choose based on their non-yield attributes) have opted for the three varieties, and 15.6 percent of them have been indifferent. When farmers were given two options (See Question 22.2, Appendix V), 18.4 percent of the respondents have opted for a single variety (looking for better yield and better market) and 81.6 percent of them have preference for three varieties (looking for more yield stability and less damaging-harvest during bad season). Thus, provided that farmers are able to get more stable yield from a diverse set of varieties, they are prepared to sacrifice extra yield from better yielding varieties.

When farmers were asked if they want to always keep local varieties, 70.7 percent of them have responded affirmatively. Their reasons were longer age, adaptability to local conditions, heritage value, less demand for intensive care, failure to trust new seeds and possibility to get some yield whatsoever the weather could be. Those farmers with no interest to keep the

FVs any more are refraining due to its low yield and lack of disease tolerance.

Coffee is a perennial crop and farmers' incentive to replace indigenous varieties by the more genetically uniform improved seedlings is mainly a function of their capacity to survive until the first harvest is ready. This is especially true for those farm households coffee is the single most income source. In most of the cases, farmers never replace all coffee trees at once. The replacement takes place over a longer time horizon than just in one year as it may be the case for sorghum. Price collapse is important to switch from coffee farming to other enterprises and not so important to switch from one coffee variety to another. Why and under what conditions do farmers decide to switch to new varieties? Table 3-1 summarizes their reasons.

Table 3-1. Farmers' reasons to switch to a new variety: coffee

Reason	Freq.
When the local variety gets old and dried	39
When there comes a new variety that is better in yield and/or when the existing one deteriorates in yield	36
When it becomes susceptible to CBDs (especially during the rainy season) and/or we get a variety better resistant to CBDs	16
Advice, education, information and encouragement from the development agent on how to plant new seedlings	7
When we get encouragement from friends and relatives; demonstration effect from other neighboring farmers	6
When the price goes down	5

Source: 2001/2002 original survey data. The source for all the subsequent tables is the same unless mentioned otherwise.

Note: The frequency column may not add up to the number of farmers due to multiple reasons by a single farmer.

Farmers have stated that a new variety unknown to them could be risky. Most do not trust a variety that they do not know. But still they would like to try a new variety to understand its behavior and compare it with the ones they know. Obviously, the farmers who are not benefiting from the varieties they already know would like to take a new variety more readily.

Yield loss (mainly due to disease), price collapse, absence of market, and lack of access to inputs (chemicals and farm implements) are the major problems for most coffee growing farmers. In ranking their production constraints, working capital, land shortage, pests and diseases have been the most important ones.

Understanding farmers' preferences for variety attributes is equally relevant for breeding priority setting as well as for on-farm conservation. To get insight on the relative importance of coffee variety attributes, farm household heads were requested to rank the importance of different variety attributes to their use and replacement decisions. Figure 3-1 below shows the results.

Figure 3-1. Ranking the role of variety attributes in farmers' use and replacement decisions: coffee

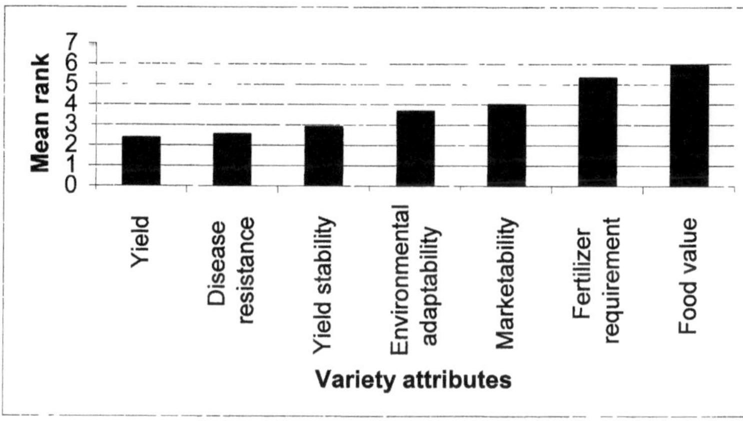

Notes: 1- Highest rank; 7 – Lowest rank. Number of respondents ranges from 128 to 257.

Lower value of the mean rank means higher importance of the variety attribute in use and replacement decisions. The figures should be conceived as ordinal values. Compared to yield, disease resistance, and yield stability, fertilizer responsiveness and food value (drinking quality of the coffee variety) get lower rank by the coffee growing farmers. The relative importance of different variety attributes on maintaining improved coffee trees and their relative importance to the household also follow the same pattern.

The importance of variety attributes could differ among farmers based on factors like prevalence of different stress factors in the locality (pests, diseases, drought, *etc.*), proportion of the crop sold and consumed, the relative importance of the crop to the household, farmers' resource endowment (land and labor) and their sensitivity to income shocks. The most important production problems, all of which resulting in low yield, are reflected in farmers' variety choice behavior. Varieties that embed the attributes with higher rank (like yield, disease resistance, yield stability and environmental adaptability[2]) will have higher probability of survival on farmers' fields and such varieties need less on-farm conservation incentives.

3.5.2. Sorghum variety choice and land use decisions

The survey results show that shortage of working capital, lack of adaptable improved seeds, weather, labor and land shortage are the five most important production constraints for sorghum farmers. Land is still a problem from many angles, namely, ownership security, shortage and degradation. The quality of the land is deteriorating and fertilizer is becoming a necessity in most settings (Negussie and Mulat, 2003). Land fragmentation and landlessness seem to be emerging challenges since farmers have to share their land with newly married children.

Farmers rank the importance of sorghum variety attributes to the household in the order yield, adaptability to environmental stress, yield stability, disease resistance and fertilizer responsiveness. About 88 percent of the sorghum growing respondents have shown interest to have as many varieties as possible. Lack of adequate land, demand for land heterogeneity by the multiple varieties, and envisaged difficulty in management are the most important reasons for the rejection of having many varieties.

Farmers who want to maintain as many varieties as possible were mainly attracted by the chance it gives them to make a choice and to make use of the different desirable attributes of the varieties. By trying more diverse number of varieties, they want to search for a variety that best fits their endowments and constraints. They want to take the opportunity that if a certain variety is bad in one attribute, the other could be good with respect to either yield, disease resistance, marketability, drought resistance, and/or food value. The diversity is believed to enable them to cope up with vagaries of nature.

When farmers were given two options (See Question 22.0, Appendix V), 68 percent of the farmers have opted for one variety. The most important reasons for this choice are simplicity for management requiring the

[2]Environmental adaptability refers to the capacity of the variety to survive difficulties like drought, birds, wind, and frost.

application of cultural practices at the same time (harvesting and weeding),[3] better market price, and better food taste. If farmers plant different varieties and harvest them mixed, they get lower price especially if the color differs. This is also true for coffee. In this case, the difference is also reflected in bean size. Those who opted for three varieties (32 percent) are mainly attracted by the possibility to try and choose among the three for other non-yield attributes (drought tolerance, pest resistance, food taste, livestock feed, and yield stability). Some farmers have also mentioned that mixing different varieties results in good food taste.

When farmers were given other two options (See Question 22.2, Appendix V), only 28.4 percent of the respondents have opted for a single variety. These farmers are mainly concerned with the yield during good season, better market price and land constraints. The majority of the respondents (71.6 percent) have opted for three varieties for either of the following reasons. Either they are living in more drought prone areas or have concern during the bad weather scenario. Still others have concern at least for children or are afraid of losing every thing under any potential stress and/or are looking for yield stability. Most of the discussions with farmers suggest that farmers with many children are more risk cautious. The results also imply that provided farmers are able to get more stable yield from a diverse set of varieties, they are prepared to sacrifice extra yield from better yielding varieties.

In the sample, 78.6 percent of the farmers want to always keep local varieties on-farm. Useful stalks (for animal feed, construction, and fuel wood), food taste, and lack of exposure with other varieties are the major reasons to stick to local varieties. Those farmers with no interest to keep the local varieties any more are stressing their yield deterioration, lack of disease and drought resistance. To get additional insights on their decisions to switch to new varieties, information was gathered on their reasons to switch. The results are summarized in Table 3-2.

[3]If different types of sorghum varieties are planted together and they mature at different times, the harvest period will have to be late until the last variety gets ready.

Table 3-2. Farmers' reasons to switch to a new variety: sorghum

Reason	Freq.
When the new variety is better on average in all attributes	39
Whenever we get a better yielding variety and no good yield from the old ones	22
When the new variety gives us good yield repeatedly (yield stability)	16
When we get good variety with a reasonable price	14
We will not switch completely because we never know	12
When the old variety no more resists disease and bad weather	8
When we get a variety that matures early	6

Note: The frequency column may not add to the number of farmers as their reasons were grouped and a single farmer was allowed to give more than one reason

Farmers will switch to new sorghum varieties provided that the new varieties are desirable with respect to the most important attributes mentioned above. The relative importance of the attributes is thus very crucial to make their decisions. Figure 3-2 below summarizes the relative importance of different variety attributes on variety use and replacement decisions.

Figure 3-2. Ranking the role of variety attributes in farmers' use and replacement decisions: sorghum

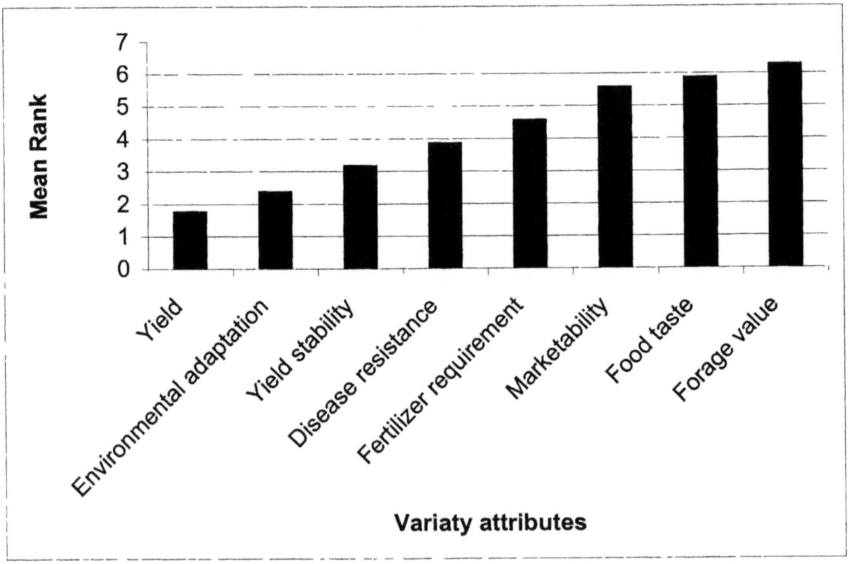

Notes: 1- Highest rank; 8 – Lowest rank. Number of respondents ranges from 147 to 185.

The importance of variety attributes could differ among farmers based on factors like livestock ownership, prevalence of different stress factors in the locality (pests, diseases, drought, *etc.*), proportion of the crop sold and consumed, the relative importance of the crop to the household and farmers' sensitivity to potential income shocks. Varieties that embed the attributes with higher rank (like yield, environmental adaptation, yield stability and disease resistance) will have higher probability of survival on-farm implying that policy makers need not design incentives for those varieties as they are maintained *de facto*.

3.6. Farmers' comparative assessment of improved versus local varieties

As far as variety use is concerned, most of the research emphasis is on the effect of agricultural technologies (like IVs and fertilizer) on agricultural productivity and farmers' livelihood. Not much is known about the desirable features of the respective varieties. Very little attention is given in the agricultural economics literature to the indigenous technology itself particularly to any advantageous attributes that the household may consider important in its cultivation decision (Meng, 1997). On top of their well known value as genetic resources, FVs are indigenous technologies by their own right. Thus, they deserve close examination not only for on-farm conservation purpose but also for enhancing their productivity in areas where they have better comparative advantage.

How do farmers perceive local and improved varieties of coffee? Critical examination of farmers' perception will play a key role to integrate on-farm conservation with utilization of crop diversity. Having this in mind, farmers' assessment of improved and local varieties is described.

3.6.1. Farmers' assessment of improved versus local coffee varieties

From among the sampled coffee growing farmers, 173 of them have planted both local and improved varieties, 86 have only local coffee trees and 7 have only improved coffee trees. Concerning their experience with improved seedlings, 25.6 percent of them buy improved seedlings regularly and 44.7 percent of them only once. Thus, 70.3 percent of the farmers have grown IVs during their life at least once.

The source of IVs for the majority of the coffee farmers (97.2 percent) is the MoA. The rest also get improved seeds from a friend or relative, private seed enterprises, and NGOs. For most poor farmers, it is difficult to get improved coffee seedlings because of either inability to pay or simple unavailability.

Land shortage is also blocking farmers from using newly coming seedlings to produce good yield (coffee and other crops).

When farmers had to choose between improved and local seedlings given equal chance, 73 percent of them have opted for IVs (mainly for CBDs resistance and better yield) and 18.8 percent of the farmers have opted for indigenous coffee trees (mainly for less intensive management and better adaptation to the local environment). The rest (8.2 percent) of them were interested to have both local and improved varieties (mainly for satisficing purposes). Further more, information was collected on how farmers evaluate the local and improved varieties with respect to different variety attributes. Figure 3-3 summarizes the results.

Figure 3-3. Farmers' comparison of the improved and local varieties with respect to different variety attributes: coffee

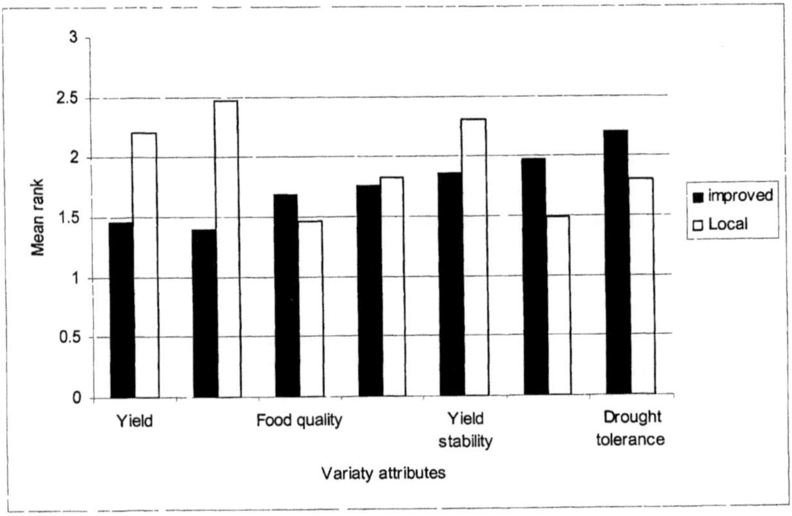

Note: 1- good; 2 - medium; and 3 - bad

On the whole, the yield and disease resistance of the local coffee trees is low but it has good food quality taste, responsiveness to fertilizer, and drought resistance. Except for disease and yield problems (which get superior rank by farmers as presented in Figure 3.1), the local variety is superior in adapting to the local conditions. Surprisingly, yield stability is better for improved coffee trees. As it is shown in Figure 3.1, yield stability is the third most important valuable attribute for coffee farmers. Based on results of

Figures 3.1 and 3.3, one can conclude that the three most important variety attributes preferred by farmers (yield, disease resistance and yield stability) are better embedded in IVs of coffee and hence on the whole most farmers will have higher propensity to plant IVs.

Coffee farmers do not unanimously agree on the yield potentials of the improved and local coffee seedlings but most have pointed out that yield is better for IVs. About 53 percent of them had given better rank to the improved coffee varieties, 8 percent to local varieties, and the rest are indifferent.

CBDs is still a serious problem for all the farmers producing coffee. Had it not been for disease, the local variety would have been good for most farmers. For marketability, most farmers (83.8 percent) are indifferent, and still the opinion of the rest is equally divided between the two varieties. This indicates that coffee farmers do not have a preference for either local or improved coffee varieties in terms of marketability. This could, mainly, be attributed to the failure of the market to discriminate among varieties. Adaptability to the local environment (drought tolerance and yield without fertilizer in Figure 3-3) is mainly associated with local coffee trees and so is disease resistance with improved ones.

For yield stability, 46 percent prefer the improved, 44 percent are indifferent, and only 10 percent prefer the local. The local coffee trees are perceived to have unstable yield due, mainly, to their susceptibility to CBDs. For food taste, 65% of the farmers are indifferent between improved and local varieties and 27.4 percent of them prefer local coffee varieties. While 40 percent are indifferent for drought tolerance, 40 percent prefer the local varieties.

It has been noted by farmers that, unlike the indigenous coffee trees, the improved coffee trees need more intensive care. The most important motive to plant the IVs is to get better yield, fast growth (early maturity) and better age. For most farmers, the reason to keep on growing local coffee trees is that it requires less intensive management (like weeding and cultivation) and it gives some yield whatever the case may be. Regarding farmers' preference for improved or local varieties during bad weather scenario, 57 farm household heads chose IVs and 47 of them opted for the local varieties. Their respective reasons are summarized in Figure 3-4.

Figure 3-4. Farmers' reasons to choose improved or local varieties of coffee during bad weather

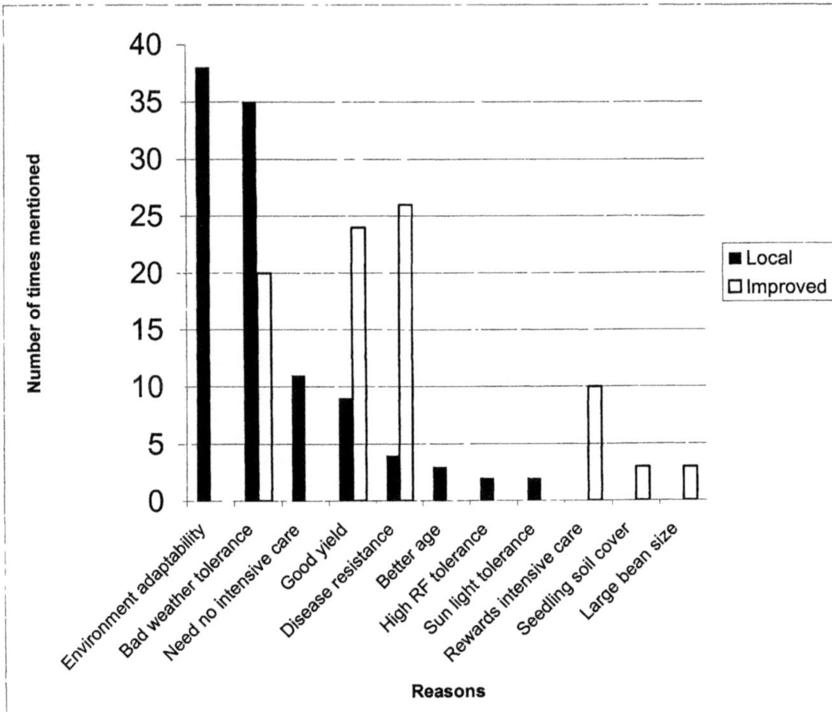

Note: The frequency column may not add to the number of farmers as their reasons were grouped and a single farmer was allowed to give more than one reason.

Mostly agricultural economists tend to think that better quality plots[4] are allocated to modern technologies like improved seeds. This claim is based on the expectation that farmers will allocate their preferred inputs to those enterprises that are most rewarding (like IVs). Accordingly, one could expect farmers to allocate the better quality plots to IVs presuming that the marginal contribution of this plot will be better with these technologies. Table 3-3 suggests that this is not the case.

[4]Bad and good quality plot is farmers' classification of their own plots based on their own criteria of ranking the fertility of their plots. Precise definition is not important here as the point of interest is to find out if farmers self-select their better quality plots to IVs.

Table 3-3: Land quality allocation between improved and local coffee trees

Farmers' groups	Freq.
Farmers who allocate equal quality plots to both varieties (either because they have the same quality plots or because of intentional reasons)	119
Farmers who have planted only local or improved	83
Farmers who allocate better quality plots to improved trees	23
Farmers who allocate better quality plots to local trees	19

Farmers sometimes plant the best plot they have to the local varieties and the more infertile ones to the IVs. The most important reason for these farmers is that they do not have much confidence on the sustainable performance of the IVs. Whatever is the state of nature, losing on the bad plot is better than losing on the good one. This reason is most frequent from those farmers with less experience with IVs. In essence, these farmers are allocating the better quality plots to those varieties that have more certain prospect. On the contrary, those farmers who are planting IVs on better quality plots are basing their decision with the features of these varieties – lack of drought tolerance and poor performance without chemical or organic fertilizer. On the whole, it can be concluded that farmers do not necessarily allocate their better quality plots to the IVs. This is in contrast to previous studies. For instance, in Mexico, local maize varieties are significantly favored over improved ones in poorer-quality soils (Bellon and Taylor, 1993).

3.6.2. Farmers' assessment of improved versus local sorghum varieties

From among the sampled farmers, 104 of them have been planting both local and improved varieties during the 2001/2002 cropping season and 94 of them were growing only FVs. Some 82 percent of the sampled farm households have grown IVs at some point during their life. The major source of IVs for the sorghum farmers (86.3 percent) is the MoA. Friends, relatives, private seed suppliers, and NGOs take the remaining share. Farmers hardly mix local varieties with improved ones but usually they do plant local varieties mixed.

When farmers were given equal chance to choose between improved and local varieties, 89 farmers have opted for the improved (mainly for good yield and early maturity reasons) and 66 farmers have opted for indigenous varieties (mainly for reasons including risk caution, animal feed, fuel wood, adaptability to local environment, food taste, and some yield even if the weather is bad). The other 42 farmers were inclined to have both varieties hoping to get from one if they lose from the other and to derive as many desirable variety attributes as possible. Some are curious just to try the newly coming IVs.

Based on the sorghum breeding strategy followed, it can be hypothesized that while the IVs are preferred for yield and disease resistance, FVs are preferred for yield stability, environmental adaptability, and food taste. Figure 3-5 summarizes the results of farmers' evaluation of improved and local sorghum varieties with respect to different variety attributes.

Figure 3-5. Farmers' comparison of the improved and local varieties with respect to different variety attributes: sorghum

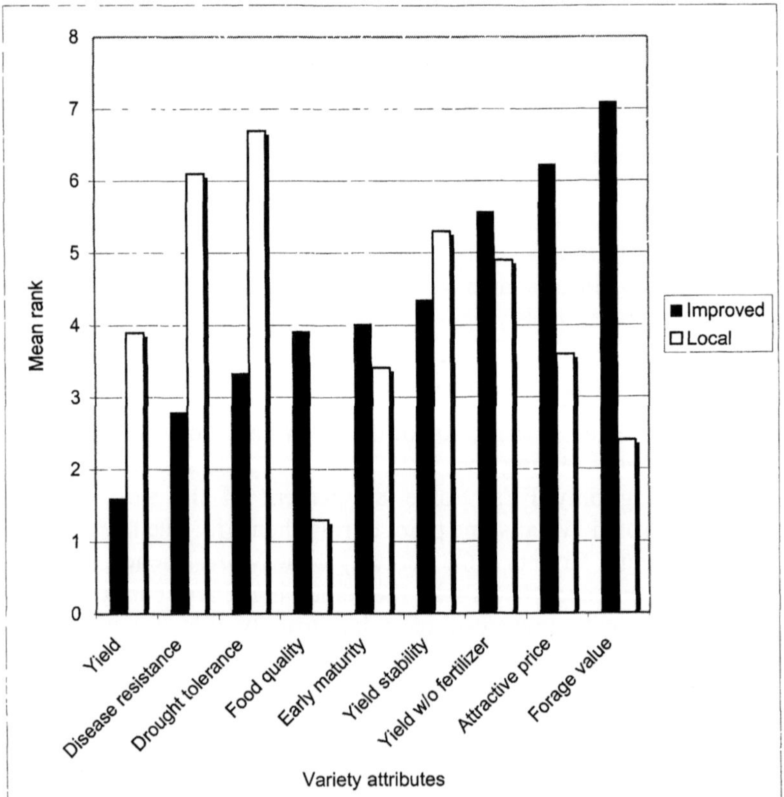

Note: 1- Highest rank; 8 – Lowest rank

All in all, the local varieties of sorghum are perceived to be very good for feed, food taste, price, early maturity, and yield stability while the IVs are good for yield, disease resistance, and drought tolerance. Most farmers prefer their indigenous varieties for fuel and construction purposes. While farmers' reasons to decide on planting local and improved varieties are similar in most respects (yield, yield stability, fertilizer responsiveness, tolerance to environmental stress, and disease resistance), marketability is more important for IVs and food taste, forage value, early maturity, and fuel use are more important for FVs. Even though the IVs at their disposal are not marketable, they would like to have IVs if they are marketable. Thus IVs are desired for the market and local seeds are mainly demanded for own consumption. If one combines results of Figures 3-2 and 3-5, it can be concluded that the IVs out-perform the FVs with respect to most of the variety attributes that have got the highest rank by farmers (yield, environmental adaptability, disease resistance, and yield stability). Since comparisons (like the one made in Table 3-2) can not tell us the equivalence of the utility of different variety attributes (for instance, how important is yield in terms of food quality?) to farmers, it is hard to make a judgment as to which set of varieties are preferred most.

Coming to farmers' preference for improved and local varieties under bad weather conditions, 89 farm household heads opted for improved varieties and 66 for local ones. Their reasons are summarized in figure 3-6 below:

Figure 3-6. Farmers' reasons to choose improved or local varieties of sorghum during bad weather

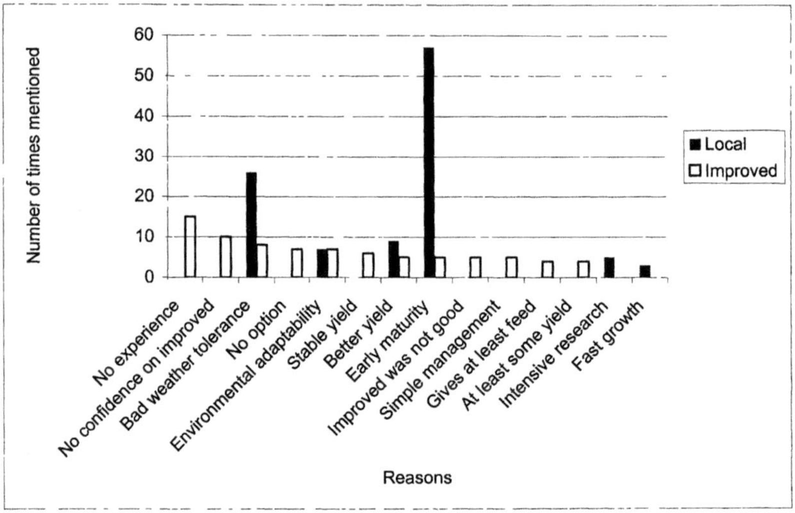

Some of the very risk averse farmers who have never used the improved variety have found it more conveneint to deal with the local varieties that are known to them and simple for management. They lack trust on the new varieties. Those farmers who have livestock prefer the local varieties because they get at least feed under worst scenario. Local varieties are preferred by most farmers for early maturity. Sorghum growing farmers give value to early maturing varieties for two major reasons. It enables them to get harvest at the critical time of food shortage and to escape water stress when rainfall is late. Some farmers attribute better utility to the IVs simply because they believe that IVs are the result of agricultural research meant for addressing their problems.

Some 42 farmers have opted for both improved and local varieties. For these farmers, the two varieties are mutually exclusive in most attributes except drought tolerance. They wish to benefit from the different desirable features of the varieties by planting both.

Last but not least, the survey was meant to see quality of land allocated to improved and local varieties of sorghum. The simple descriptive statistics in Table 3-4 discloses that most farmers who have multiple plots allocate equal quality plots to local and improved varieties either because their sorghum plots are of equal quality or they are indifferent in their allocation decisions.

Table 3-4. Land quality allocation between improved and local sorghum varieties

Farmers' group	Freq.
Farmers who use only the local varieties	94
Farmers who allocate the same quality plots for both varieties (either because they have the same quality plots or because of intentional reasons)	81
Farmers who allocate better quality plots for local varieties	14
Farmers who allocate better quality plots for improved varieties	9

As it is the case for coffee, those sorghum farmers who have allocated better quality plots for local varieties noted that they never trust IVs while those who allocated better quality plots to IVs were more confident and interested with the yield potential of the IVs.

3.7. Comparing coffee and sorghum variety characterization and choice criteria

Sorghum is not exclusively the 'poor man's crop' and neither is coffee exclusively a 'rich man's crop'. Most of the differences between coffee and sorghum variety choice behavior emanate from the nature of the growth pattern and importance of the crops in the market. Coffee is perennial and cash crop mainly produced for the international market but sorghum is annual and food crop mainly produced for food and the domestic market. Age of the variety, for instance, is not an important criterion for sorghum variety choice. Moreover, price is a major concern for coffee farmers and household use (food taste and forage value) is a major concern for sorghum farmers. Feed value is not important for coffee farmers as its chaff is not used for livestock feed. In both sorghum and coffee, most farmers are following a '*maxi-min decision strategy*' whereby they are more concerned with achieving the minimum requirement for household survival.

The average age of varieties grown by farmers is an expression of the rate of varietal replacement (Hartell *et.al*, 1998) for annual crops. For coffee, most farmers have maintained the local coffee trees (locally called '*Begeja*') for more than 20 years (average 25.4) ranging from 1 to 124. For sorghum, the variety turn-over ranges from 6 to 10 years, the average being 7.4 years. Thus, variety turn-over is higher in sorghum compared to coffee. Heritage value of maintaining FVs is typically important for coffee trees.

In the case of coffee, unlike sorghum, switching to new coffee seedlings is more difficult as it will take time to get the first output. Thus, removing old coffee trees and replacing them by new ones is done progressively to minimize the effect on household well-being. In contrast, switching from one variety to another is relatively easy for sorghum.

In general, coffee farmers hardly abandon varieties, be it local or improved. Only 16 farmers had that experience for local coffee varieties and only 3 for improved ones. The abandonment is due mainly to susceptibility to CBDs. In contrast, sorghum farmers usually abandon FVs from their fields. While 36.4 percent of the sorghum growing farmers have the experience of abandoning local varieties, 20.5 percent of them have the experience of abandoning IVs. The most important reasons for abandoning local sorghum varieties are late maturity (and hence liability to rainfall stress), poor yield, too much labor demand, bad taste for food and feed, and susceptibility to wind. For the improved sorghum varieties, the reasons for abandoning were poor yield, damaging leaves for animals and susceptibility to pests, disease, locust, and drought.

Sorghum variety use decision is influenced by suppliers *i.e.* 36.8 percent of them have reported to have been influenced by seed suppliers like the MoA. In contrast, coffee variety use decision is hardly influenced by seed suppliers.

3.8. Understanding farmers' varietal decisions for model building

What has been done in this chapter is an intermediate input to approach each of the empirical research questions. Understanding farmers' potentials, constraints, value judgments, and resource endowments will serve as an input for modeling their diversification behavior, attribute preference, and expectations. Once it is understood as to what varieties and attributes are most important and the reason(s) behind, it will be used to derive crop diversity and attribute preference as the outcomes of farmers' utility maximization. It will then be possible to predict how farmers will react once on-farm conservation schemes are in place.

When farmers shift from one variety to another, losses and gains (in terms of variety attributes) are involved. This shift will also have its impact on the survival of local varieties on farmers' fields. To harmonize the potential trade-off between crop diversity and use of modern inputs, farmers' perceptions about the improved and local varieties should be understood. Understanding farmers' preferences for the two competing technologies and what each variety attribute means to them will take the empirical modeling of opportunity costs one step ahead.

The monetary and non-monetary gains and losses of switching from one variety to another are the outcomes of variety attributes. In the long-run, policy has to move in the direction of integrating on-farm conservation with priorities identified by farmers. The information contained in this chapter will help further examine as to what crop diversity, multiple variety attributes, and FVs mean for farmers and their livelihoods.

3.9. Summary and concluding remarks

In this chapter, sampled farmers' variety choice behavior has been described. Coffee and sorghum are two important crops not only to the Ethiopian agricultural economy but also to its biological resource stock. Having this in mind, these are the crops about which data are collected and described in this chapter. Farmers' variety choice behavior was descriptively examined at length. Moreover, their perceptions about the improved and local varieties have been examined. The results by and large show that farmers mainly maintain crop variety (ies) based on their utility to their livelihood strategies.

Having said so much on the data, what are the issues involved in trying to address crop diversity conservation and economic development? How do adoption of IVs, crop diversity, and markets interact at the household and national level? Which economic theories are we using to address our empirical research questions? These are some of the points that will be taken up in the next chapter.

4. Conservation of crop diversity and economic development: issues, paradigms and economic theories

4.1. Introduction

To conserve crop diversity, policy makers have to deal with issues of crop diversity that include property rights, uncertainty, valuation, and measurement. On top of these, institutional and policy failures are still remaining challenges for most developing countries. The potential trade-off between crop diversity and diffusion of IVs or market development is another source of concern.

Severe rural poverty coupled with nature dependence of agriculture and the fragile environment where it is operating call for policy measures that could involve agricultural intensification, changes in property rights and institutions, and market development. There is a crop diversity opportunity cost when government wishes to implement policies geared to transformation of agriculture. In the light of such trade-offs and gaps, the purpose of this chapter is to present the issues involved when dealing with crop diversity research, conservation and agricultural development. In doing so, the chapter wishes to synthesize the controversial issues and challenges that potential stake-holders face.

The chapter is organized as follows. It begins with a brief review of the current conservation paradigms in Section 4.2. Issues and challenges in crop diversity conservation and research follow in Section 4.3. Section 4.4 deals with the farm level factors potentially leading to crop diversity loss. The national level factors complicating conservation of crop diversity are the subject of Section 4.5. The economic theories relevant to conceptualize the objectives of the thesis are presented in Section 4.6. Section 4.7 provides a brief summary.

4.2. Conservation of crop diversity: current paradigms

How much biodiversity does society need? Basically, there are two influential views (*utilitarian* and *ethical*) as to the conservation of biodiversity in general and crop diversity in particular. While proponents of the *utilitarian* view emphasize the *use value* of crop diversity, the supporters of the *ethical* view emphasize the *existence value* of crop diversity. The *utilitarian view*, accepted by most economists, asserts that crop diversity has to be conserved based on its marginal utility to society. The most immediate reasons for caring about biodiversity in economics are *instrumental* and *utilitarian* (Barbier, *et.al*, 1994). On the contrary, the *ethical view* claims that all biological resources are valuable and should be conserved for *moral* and *ethical* reasons (Fox, 1984). Unlike the *utilitarian* approach, *moral* and *ethical* reasons for conservation do not lend themselves to calculations of exact financial value (Maxted *et.al*, 1997). In the presence of these differences, however, the ethical and utilitarian reasons for conservation do also overlap (Hawkes *et.al*, 2000).

Both economists and biologists acknowledge that crop diversity is valuable. A question on which economists might differ from biologists is how many species should be conserved (Evenson *et.al*, 1998). To the economist, costs of conserving necessitate some implicit ranking of genetic resources based on their value. The resource allocation decision in economics is guided by anthropocentric values of resources (Bhattarai and Hammig, 1998).

From a utilitarian perspective, conservation cannot be regarded as an objective in itself. Crop diversity conservation should be undertaken to the ultimate goal of serving human kind. That is why the Convention on Biological Diversity (CBD), among other things, is concerned with not only conservation but also the sustainable use of biodiversity for development (Hawtin and Hodgkin, 1997).

Of course, it is not possible to preserve every thing (Weitzman, 1992). For economists, it is neither meaningful nor practical to deal with conservation of everything. However, given all the uncertainties involved about the costs of losing and benefits of conserving biological diversity, a concept like 'the most valuable species' is generally problematic (Weitzman, 1992). This is because of the fact that the traditional economic value of a species is of no significance in determining its usefulness as an environmental indicator (Ehrenfeld, 1976).

The approach that says every life form must be preserved tends to deny the possibility of setting priorities (Tisdell, 1990). As it will be made clear subsequently, the thesis generally follows the utilitarian approach and pinpoints the need for priority setting, identifying crop varieties conserved *de facto* and those that require on-farm conservation schemes. To face the challenge in priority setting, there is a need to first develop an '*early warning system*' after identifying '*endangered species*' and '*hot-spots*'.

4.3. Crop diversity: issues and challenges in research and conservation

The structure of property rights, the public good nature of the resource, the fundamental uncertainty about the implications of species loss, and the distortionary effects of poverty are common factors in the over-exploitation of biotic resources (Perrings and Opschoor, 1994). There are different complexities facing biodiversity research and conservation policy. These difficulties are summarized here so that subsequent arguments, methods and results can be seen in the context of these debates.

4.3.1. Difficulties of disentangling public and private aspects of crop diversity

Unlike most other resources, seed is two things in one (Morris and Heisey, 1998; Sedjo, 1992; Mountford and Keppler, 1999). First, it is a consumable input, *i.e.* something that can be combined with other inputs (e.g. land, labor, water, fertilizer, pesticide) to produce a crop. Secondly, it is also a source of germplasm, *i.e.* a store of genetically encoded information that determines how other inputs combine and become transformed into useful products. When a farmer chooses to plant a variety or combination of varieties based on their observable characteristics, s/he also chooses to use an unobservable set of genetic information along with their attributes.

When FVs are used in public breeding activities, they are public goods with social value (Brush and Meng, 1998). Moreover, the information embodied in a plant material can be replicated at low cost (Evenson *et.al*, 1998). This is their *non-rivalry* character (Simpson *et.al*, 1996).

The functions of crop diversity have public goods dimension beyond their territory – reduced probability of disease outbreak, increased productivity that benefits consumers and ecological *resilience*. All these benefits are consumed by any one in the society *i.e.* as genetic resources, the benefits of maintaining FVs are *non-excludable*. Conservation of biological resources is a public good not only from the stand point of the beneficiaries but also from the stand point of the species themselves (Perrings *et.al*, 1995) *i.e.* it is not possible to protect one species without protecting others which share the same habitat. Thus, as genetic resources, local seeds are non-excludable and non-rivalry. To the farmers, considering them as consumption goods, seeds are also private goods *i.e. excludable* and *rivalry*.

Not only the benefits of crop diversity are of *private* and *public goods* nature but also the problems associated with loss of crop diversity are *private* and *public bads*. A farmer suffering from crop failure due to lack of tolerance of the variety (ies) is facing a private bad. Other members of the society are also facing problems of food shortage, hunger and poverty.

In practice, conserved genetic resources are taken as public goods at local, national and international levels (Hawtin and Hodgkin, 1997; UNEP, 1994). For any informed policy decision, identifying the *public* and *private* benefit(s) of conserving biodiversity is as valuable as identifying the *public* and *private* cost(s) of losing biodiversity. However, it is not so simple to exhaustively list the benefits of conserving or costs of losing them let alone to completely group them either as public or private goods/bads. The problem of disentangling public and private aspects of genetic resources makes it difficult to delineate who is benefiting from biodiversity (the agent that has to pay for conservation) and who is contributing for conservation (the agent that has to be rewarded). Despite the problem of disentangling, public goods nature of crop diversity makes it by and large a public concern. That is why biological diversity is described as a 'common concern of human kind' (UNEP, 2000).

4.3.2. Property rights issues

Questions like: 'Is it possible to properly define private property rights for genetic resources?' 'Is it technically feasible?' and 'Is it, after all, important to define property rights in crop diversity?' have no simple answers. Assigning property rights for genetic resources is particularly challenging (OECD, 1997, 1998; McNeely and Vorheis, 2000). Patenting local varieties, for example, is not possible because a patent must meet the criteria of novelty that is difficult to certify for local varieties (Virchow, 1999a). Naturally occurring plants or animals are not considered patentable inventions (Frisvold and Condon, 1998). Thus, crop diversity is very difficult to break down into discrete commodities that fit neatly into a property rights system. This will make it difficult to identify the contributors and the users.

Given this situation, it is not clear what is meant by farmers' rights. Usually, scientists argue in favor of respecting farmers' rights (Swaminathan, 1996) and valuing their contribution in maintaining diversity of multiple crops along with the indigenous knowledge. However, operational implementation of what ought to be done will prove to be complex to say the least. As gifts of nature, every one in the economy faces the benefits of conserving crop diversity and the costs of losing it.

Difficulty of assigning property rights for crop diversity is commonly cited as one of the institutional hurdles in the conservation of biological resources (Virchow, 1999a). How can this difficulty be tackled? To make property rights operational in genetic resources, there is a need to redefine or change the patenting criteria (Swanson and Göschl, 2000b).

In this thesis, we argue that any attempt to try to assign property rights of crop genetic resources to individual farmers or farmers' groups within a country is a futile exercise. However, a country can claim the right to use or sell its genetic resources to any country on demand. Due to the difficulty of assigning property rights of these resources for individual farmers, in this thesis, the following property rights scheme is proposed. The scheme works within a nation and at the international level.

To its practical implementation, we propose the classification of benefits of biodiversity conservation in the following way. There are *benefits demanded* and *benefits enjoyed de facto*. Benefits enjoyed *de facto* are those benefits of biodiversity conservation that go beyond the boundary of the conserving agent at national and international level without the intention of either the supplier or the beneficiary. The benefits of biodiversity conservation are diffused across international boundaries (Perrings, 1995). For instance, take the ecological benefits of maintaining biodiversity. For these kinds of benefits, it is not possible to force the beneficiary pay as it is not the case for farmers to get paid for their *de facto* conservation.

On the other hand, *demanded benefits* are benefits that the resource user demands either through payment or by harvesting without payment. The benefits of harvesting from biodiversity are captured by the individual resource user (Perrings, 1995). For demanded benefits, policy can design and enforce payment schemes based on the value of the derived output and the time horizon during which the return from the product (pharmaceutical products or IVs) flows.

At the national level, when government or private agencies wish to use a plant genetic resource conserved *de facto*, they can use it for free. Demanded benefits of *de facto* conserved genetic resources can be earned freely at least until positive externalities *of de facto* conservation are internalized. When demanded beneficiaries wish to use a genetic material conserved (in situ or ex situ) at cost, they have to cover at least the cost and agree to pay back portion of the benefit they derive.

International users of a country's genetic resources (conserved *de facto*) should reach an agreement with suppliers to pay part of the benefit. For instance, the Californian example presented in Section 1.2 is a good practical case. This requires designing enforcement mechanisms for international treaties like the CBD. International users of any genetic resources (conserved at cost) should take part in the cost of conservation, pay an agreed amount at the delivery, and pay part of the benefit for an agreed number of years. Given this, the possible property rights regime that can hold is that a donor country is given the right to get compensation once a valuable product of commercial value is produced from a germplasm taken from its genetic stock. In effect, a source country trades the right to collection in return for a guarantee of some portions of the revenues generated by commercial development of a product that utilizes a genetic constituent from a wild genetic resource collected within the country (Sedjo and Simpson, 1995).

4.3.3. Issues in the valuation of crop diversity

To take care of crop diversity, society should value it. However, valuing genetic resources and estimation of the marginal benefit of conserving each type of genetic material is challenging because identifying the appropriate contribution of the use of conserved germplasm to the gains in plant breeding or pharmaceuticals is difficult, if not intractable. Identifying the contribution of a genetic material to the success of a particular improved variety is a complex task (Pearce and Moran, 1994) and so is the identification of the returns to genetic materials net of complementary inputs. The valuation problem is related to the problem of uncertainty as policy makers do not know the type of attributes required, the time at which they are required, and the quantity of the genes required (Virchow, 1999a).

Given the difficulties in the process, one could argue that valuation of crop diversity is simply an academic exercise as the values can never be fully captured using monetary figures. It is always a question of *second best i.e.* having some figures is better than nothing. Due to all these complications, some authors have pointed out that valuation of biodiversity is a non-sense and a hopeless exercise (Ehrenfeld, 1988). Because of the impossibility of capturing the true value to society of crop diversity, the valuation of benefits from conservation of crop diversity is no more than an indication of the minimum national incentive that the country has for undertaking any conservation effort. If it is possible to justify that the minimum benefit is larger than the cost, there is no need for a complete valuation of benefits (Pardey *et.al*, 1999; Koo and Wright, 1999). Once the minimum benefit expected from conserving crop diversity is set, valuation can help in priority setting and cost sharing.

Figure 4-1: Agricultural value of crop diversity and the need for valuation

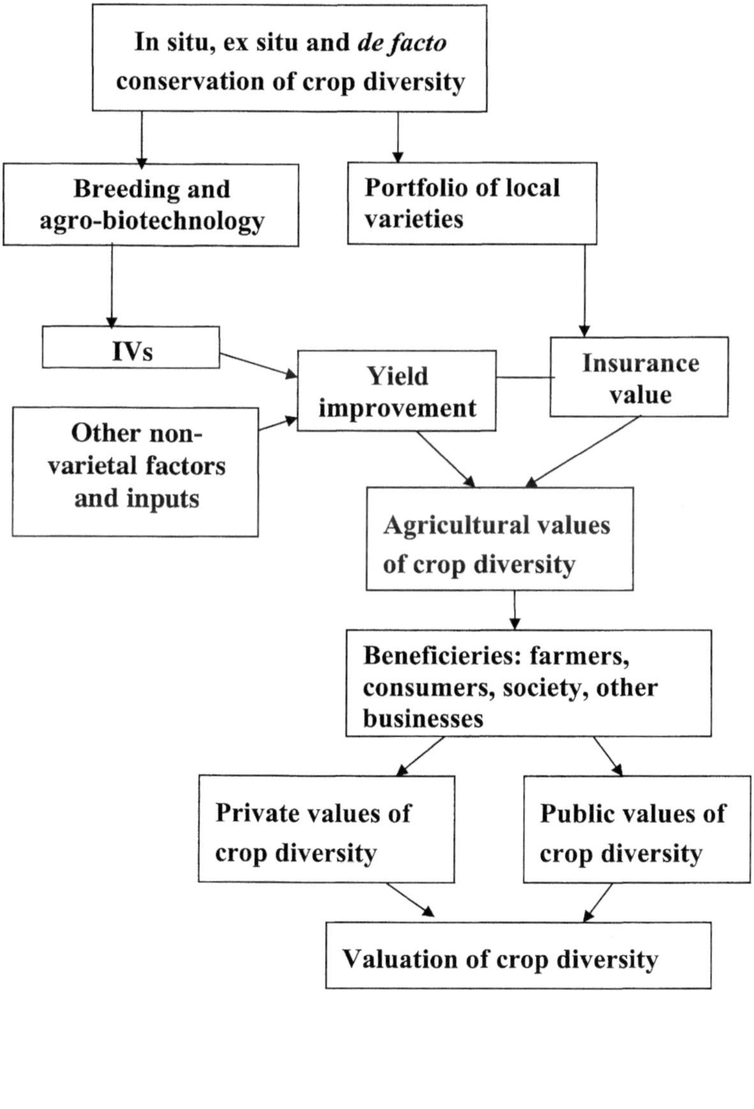

Figure 4-1 above sketches the agricultural value of crop diversity and motivates the need for valuation. As noted in Section 1.2, crop diversity has two major utilities in agriculture: *productivity value* and *insurance value*. Maintaining crop diversity offers a chance to stabilize yield and is a means to get hybrids or better varieties adaptable for certain production constraints. These two constitute agricultural values of crop diversity. The agricultural values of crop diversity are valued differently by the private individuals and the public due to market failure and public/private aspects of crop diversity. This divergence necessitates the need for valuation of crop diversity.

It has to be noted that the use value of the marginal unit of biodiversity as an input in pharmaceuticals and related products is much smaller than the opportunity or protection costs of the lands on which biodiversity habitat resides (Sedjo and Simpson, 1995). The financial values of genetic resources could be quite low due to their redundancy.

4.3.4. Uncertainty issues

What types of genetically coded information will be required? When, where and how frequent will they be required? There is a tremendous uncertainty on these questions because of the impossibility to predict what types of agricultural production problems will have to be tackled.

The uncertainties surrounding the over-exploitation of natural resources in general and biodiversity in particular, and the problem of irreversibility have long been subjects and concerns in natural resource economics (for instance, Krutila, 1967; Arrow and Fisher, 1974; Segerson, 1992; Pindyck, 1991). The problem is not that private decisions on the use of biological resources involve externalities but that those externalities may involve irreversible loss of resilience in ecological systems that provide essential life support services (Perrings *et.al*, 1995; Simpson *et.al*, 1996).

The damage may not even be noticed by the nation or the international community until it is too late to do anything to avoid permanent degradation (McNeely and Vorheis, 2000). Accordingly, expected payoffs from productive investments differ from expected losses from failure to make defensive investments (Bromley, 1992). While it could be possible to prevent biodiversity loss in the immediate present, huge costs can be called for once impacts start to result in concrete losses. Lack of complete information on the benefits of conserving and costs of loss always hinders the success of biodiversity conservation and utilization.

In short, it is not possible to accurately predict what may be useful in tomorrow's world (Smale, 1997). Of course, the best thing to do regarding risk in biodiversity is to attach a probability to the occurrence of different states of nature and to base the decision on the diversity index constructed. The *precautionary principle* in this connection suggests that where there remains some chance of serious and irreversible damage, preventive measures should be taken even if the science is insufficiently developed to assign reliable probabilities to the occurrence or magnitude of the damage (Perrings and Opschoor, 1994).

The other issue that biodiversity research and policy is uncertain about is the question of loss itself. Some studies question the mere existence of biodiversity loss or extinction. The controversies range from questioning the existence of the phenomenon of genetic erosion itself (Smale, 1997) to taking it for granted (Perrings and Opschoor, 1994). If species are not discovered in the future, one cannot be sure whether they became extinct or never existed in the first place (Mann, 1991). As it is hard to know the total stock available at a point in time, it has been hard to prove if at all extinction or loss occurs.

To sum up, the benefits of crop diversity are not immediate and their need is unpredictable. At times, there may be a need to invest on the conservation of these resources with no immediate use value. Because of the uncertainty on the problems agriculture is going to face (drought, pests, insects, disease, *etc.*) in the foreseeable future, there is always a need to maintain genetic diversity of future potential agricultural value. The *pre-cautionary principle* presented above also supports this need.

4.3.5. Issues of crop diversity measurement

Measuring diversity itself is one of the challenges in the crop diversity research. Biodiversity is such a complex entity that it cannot be measured and expressed as a single quantitative variable in the way that, for example, length, area, and mass can be (Perlman and Adelson, 1997). No one measure of biodiversity alone will help conservation organizations as they articulate and set priorities for conservation.

What options are there to measure on-farm crop diversity? There is a growing literature on the economic theory of diversity attempting to measure and quantify its value (Weitzman, 1992, 1993; Polasky and Solow, 1995; Solow *et.al*, 1993 and Nehring and Puppe, 2002). All these studies have the common problem of attaching subjective weights to the different components of biological diversity.

Diversity theory can be viewed as a branch of utility theory with a focus on how marginal utility depends on the set of goods consumed (Nehring and Puppe, 2002). Diversity theory is all about what and how much diversity to conserve. The 'how to conserve?' question is a technological question for which diversity theory does not help much (Puppe, 2003, personal communication).

In the research on the economics of crop diversity, variety count, ShaNNon's, and Simpson's diversity indices are the most commonly used diversity indicators (Smale *et.al*, 2001; Meng, 1997; van Dusen, 2000). For our purpose, count of FVs as named by the farmers themselves, number of variety attributes important to the farmers' livelihoods (attribute count), and proportion of land allocated to local varieties of crops are the indices used in chapters 5 and 6 to represent crop diversity. Attribute count is a new diversity index proposed. The added values of attribute count and its features will be explained in Chapter 6.

As far as crop diversity is concerned, the function that society would like to maximize is diversity at the national level (D_n) such that it contains as many distinct attributes as possible and as evenly distributed as possible. The problem of the indices like Count index, however, is that they assume that all FVs contribute equally to crop diversity This is not the case because the distribution of each also matters (Magurran, 1988). Therefore, there is a need to consider frequency of individual occurrences. To this end, there are many diversity indices (richness, evenness and dominance)[1] suggested in the ecology literature (Magurran, 1988). In this thesis, ShaNNon's diversity index is also used. Encompassing both richness and relative abundance, it is defined as (Magurran, 1988):

$$D_s = -\sum_{i=1}^{s}\left[\left(\frac{n_i}{n}\right)\ln\left(\frac{n_i}{n}\right)\right] \quad \text{ShaNNon's diversity index (4.1)}$$

[1]Diversity indices could be constructed for purposes of richness (the number of species encountered in a given sampling effort), evenness (equality of distribution of abundance) and dominance (the distribution of individuals associated with each of the species) (Magurran, 1988). For our purpose, richness refers to the number of FVs of a crop per farm. Evenness refers to the extent to which the different varieties are planted evenly on each farm. Dominance refers to the extent to which either of the varieties dominates.

where n_i is the number of individuals belonging to the ith of 's' species in the sample and n is the total number of individuals in the sample. For our empirical analysis, species is proxied by farmers' named varieties and n_i/n by the area share occupied by the ith variety[2]. ShaNNon's index could be divided by ln's' to get evenness of the varieties where 's' refers to the maximum crop diversity when all species are equally abundant which is the number of FVs counted on each farm.

If one expects that each local variety is contributing equally to aggregate national diversity and each one is equally valuable and equally distributed, variety count could suffice. If one believes that the more the share of variety A, the more important it will be to the farmer and to the national crop diversity, then one can consider ShaNNon's diversity index as a diversity indicator. For this thesis, the ShaNNon's indices have been used to see the extent to which the count diversity estimation results are sensitive to evenness and relative abundance (See Tables III-5, IV-3, and IV-4 in the Appendix). Most of our discussions (in Chapters 5 and 6) and policy implications derived from the results (Chapter 9) are, however, based on the variety count, proportion of land and attribute count indices[3].

4.4. Potential farm level factors leading to crop diversity loss

4.4.1. Potential 'replacement' by hybrids and modified varieties

Genetic uniformity because of agricultural development and environmental change has been a concern for practitioners during the last four decades. This concern is motivated by the potential 'replacement' of FVs by a few genetically uniform IVs (Brush, 1991). It was further fueled by the green revolution that has transformed agriculture in many countries. However, the impact of diffusion of IVs on crop diversity is one of the contentious issues in the crop diversity literature. Even though the controversy needs to be settled empirically, in general, there are two polarized views.

---

[2]When farmers could not tell their variety-wise land allocation as they often mix the varieties, they were asked to rank based on the amount of seed they think they have planted from each type of variety. When it was said equal, the land allocated to the crop is apportioned equally among the reported varieties.

[3]For the purpose of long-term on farm conservation decisions, variety and attribute count indices are more important than the other indices considered. This is mainly because the proportion of the area allocated is a one year decision that is made during the year the survey was made, 1992/93 Ethiopian Calendar cropping season, especially for sorghum.

While one strand of literature asserts that replacement of the FVs by genetically uniform IVs is threatening crop diversity (Smith, 1969; Brush *et.al*, 1992; Bellon *et.al*, 1997) the other strand takes IVs as having important role in maintaining and even enriching genetic diversity of indigenous cropping systems (Wood and Lenne, 1997). Swanson (1999) also has considered adoption of IVs as the cost of development and accordingly there is a clear opportunity cost of agricultural development interventions such as improved variety dissemination. According to Brown (1983), the first criticism against adoption of IVs, genetic erosion, is a perfectly valid one with which no responsible biologist should disagree.

Genetic erosion model is mainly challenged by results of the partial technology adoption literature (Just and Zilberman, 1983; Feder *et.al*, 1985; and Feder and Umali, 1993). Accordingly, farmers do not adopt a technology (be it improved variety, fertilizer or method of input application) in its entirety with all its components but only part of it depending on the conditions they live in. *Asset specificity*, portfolio selection, safety-first behavior and learning are the four major competing explanations that microeconomic theory offers to partial adoption of improved technologies (Smale *et.al*, 1994). For instance, to learn more about the technology, the early adopting farmer, unlike the late adopter, may choose to adopt a component of the package rather than the complete package (Feder and Umali, 1993).

There are implicit assumptions behind the presumption that IVs replace indigenous varieties (Perales *et.al*, 1998). It is presumed that IVs are always superior to FVs in economic profitability and all farmers share the objective of maximizing expected profits. However, IVs may not have clear advantage over FVs in terms of yield or net income per hectare (Perales *et.al*, 1998; Edilegnaw, 2003) and there is no need to necessarily adopt IVs to improve productivity (Brush *et.al*, 1988). A local variety that is better adapted to local agro-climatic conditions may be even more successful than its improved counterpart on a plot of low fertility with no irrigation. Above all, farmers' objective is far more diverse than just maximizing farm income.

In areas where there are no adequate markets for agricultural products, increased farm yield does not translate to increased income as a result of inelastic demand for agricultural products. The drop in the price of agricultural products can partly bring about dis-incentives for farmers to adopt IVs.

Most studies show that farmers who adopt IVs frequently also continue to plant FVs and complete displacement does not occur (Brush *et.al*, 1992; Brush, 1986). The important policy question concerning the issue of potential replacement is finding the right mix of policies to maintain FVs and improve agricultural productivity. Both sets of varieties have their own advantages and disadvantages (both to the farmers and the nation at large). Thus, policy can not favor one set of varieties at the expense of the other set. Optimizing use of local and improved varieties without disturbing the genetic stock should be the objective function for policy makers.

The genetic diversity implications of adoption of improved varieties of crops will remain to be the most important controversial policy issue in the development literature. It is yet not clear whether it is addition or substitution which is taking place in reality. The model that predicts crop diversity loss as a result of diffusion of IVs is, thus, highly controversial not only because there is no clear-cut replacement but also the impact of using IVs on crop diversity is not known with certainty. In some cases, the IVs are not hybrids but just simple selections from among the available FVs. Chapters 5 and 6 will examine the long-term impact of experience in using IVs on coffee and sorghum diversity on-farm.

4.4.2. Access to markets and on-farm crop diversity

Access to agricultural markets and related improvements in rural infrastructure and marketing institutions are essential for adoption of new technology and transformation of subsistence-oriented smallholder agriculture (Zeller *et.al*, 1998). However, such agricultural development interventions, despite their positive impact on rural household incomes, may have crop diversity opportunity costs.

In the absence of markets, farmers attain their household objectives autarkically (de Janvry *et.al*, 1991). If this is the case, farmers have to find their own internal mechanisms of dealing with any household needs. One of these mechanisms is managing wider portfolio of crops or varieties under cultivation that can give a better chance to the farmers to satisfy their diverse consumption and production needs and protect themselves against the vagaries of nature. On the contrary, economic theory suggests that farmers in the more market integrated localities have a better chance to use the market to buy what they need and sell their surplus. In this case, the market is a balancing mechanism that absorbs the surplus and fills any gaps and self-sufficiency in varieties and variety attributes may not be important.

Producing more by using modern technologies like IVs and fertilizer can be more attractive in the second scenario. Thus, on the whole, economic theory predicts negative impact of access to markets on crop diversity. The theoretical models derived in Chapters 5 and 6 will empirically test the impact of access to markets on crop diversity.

4.5. National level factors complicating crop diversity conservation

The underlying causes for resource degradation in general and crop diversity loss in particular in developing countries could be attributed to policy failures, market failures, and institutional failures to mitigate the impact of development interventions. Poverty at the national level will deprive the incentive for the country to undertake conservation of its crop diversity by limiting its capacity. National poverty emerges as a challenge because poverty reduction strategies appear to compete on public investment strategies geared to conservation. Moreover, there are trade-offs involved between agricultural development interventions and their consequence on crop diversity. However, as it will be shown in Chapters 5 and 6, poverty at the household level is resulting in crop diversity production *de facto*[4].

4.5.1. The challenges of facing development and conservation of crop diversity

Knowledge, research and development infrastructure (status of crop improvement program, the participation of the private sector, and information flow), and financial capacity are the major assets for a country to sustainably conserve and utilize its crop genetic resources.

Economic development and poverty reduction are the first and over-riding priorities for developing countries. The same holds true for Ethiopia. These priorities, however, can not be expected to be pro-diversity. Crop diversity conservation and poverty reduction are in most respects competing. Thus, conserving crop diversity could be technically difficult and challenging for developing countries like Ethiopia as they have over-riding poverty reduction priorities. Many of the studies on poverty-environment relationships support the notion that poverty causes increased pressure on the environment (Dasgupta, 1993, 1995, 1996).

---

[4]Thus, in situ conservation costs will increase with the increase in incomes of farmers and ex situ conservation will become more advantageous compared to in situ in the course of economic development.

Conserving indigenous crops and their genetic diversity on-farm is tantamount to trying to stop agricultural development (Brush, 2000). As the frontier of the modern sector expands, homogeneity continues to replace diversity (Swanson, 1995). The common response from policy makers to improve agricultural productivity is agricultural intensification. The effect of this move on crop diversity, according to the genetic erosion model, is still negative. On the whole, it can be said that, given the prevailing institutional set-up, registering agricultural development without degrading crop diversity will remain to be challenging to developing countries.

4.5.2. Market failures

The market fails to value most of the biological diversity. It also fails to reward those economic agents that contribute to the diversity outcome. Further more, as it is the case with other natural resources, the market does not debit and credit the state of genetic resources. As a result, national accounting information is usually generated using only those items involving financial transactions. In all these cases, the market is considering biological resources as resources of no economic value or as resources that can be extracted for free indefinitely. If the *status quo* remains as it is, the prevailing market system could fail to retain all the crop diversity that society may need because it will result in the production and marketing of only varieties embedding those attributes that get better price and household utility.

When a farmer purchases or sells 10 kilograms of sorghum in a village, it is the commodity for which he/she is paying or getting paid; not the genetic information contained there in. The farmer will maintain those crops that have some visible benefits to the household. Those varieties with important attributes (for future breeding) with no attractive features to the farmers today could be abandoned from farmers' fields. The gap between the social and private value of crop diversity demands policy interventions that internalize the social values of crop genetic resources. Such interventions should offer national incentives to undertake conservation of crop diversity.

4.5.3. Institutional and policy related constraints

Despite the presence of many international agreements and conventions (CBD signed by 168 countries, Trade Related Aspects of Intellectual Property Rights (TRIPS), and International Union for the Protection of New Varieties of Plants (UPOV), most of the actions and their effects at the grass root level are rather yet to be seen. The major problem seems to be enforcement. The problem, among other things, is aggravated by the complications discussed in Section 4.3 and Sub-sections 4.5.1 and 4.5.2 above. Institutionalizing the operational implementation of international conventions demands institutional innovations at the local, national and international levels. For the most part, the problem is not absence of local and international rules but enforcement mechanisms to institutionalize the value of crop genetic diversity to society and involve the users in the cost of conservation.

At the international level, the conflict of interest between the poor countries (net suppliers of biodiversity concerned with farmers' rights) and the rich countries (net demanders of biodiversity concerned with intellectual property rights) has to be resolved. Unlike poor countries, the rich ones have better capacity to use crop diversity. The concern for the poor countries is benefit sharing and financing conservation. Because of the capacity problem, over-riding need to address poverty, and the high concentration of already existing crop diversity, the potential for genetic diversity loss will be high in poor countries unless institutions are created to mitigate the impact of development.

Institutional incompatibility is the other most important prevailing problem. It refers to lack of coherence among institutions of different nature and scope: formal and informal, local, national and international[5], and institutions crafted for development purpose versus those for conservation purpose. There is too much lack of coordination among the different conservation and development agents. One case in point in Ethiopia is MoA's attempt to increase food production through intensive agriculture and the IBCR's attempt to undertake conservation of landraces on farmers' fields. These objectives are not necessarily compatible. There is conflict of interest among such organizations and policy makers need to clearly spell out ways and means of crafting compatible institutions harmonizing vested interests of farmers and the respective organizations.

To curb the genetic resource loss, the Ethiopian government has formulated National Policy on Biodiversity Conservation and Research in April 1998.

---

[5]Because of differences in the alignment, the lack in coherence between the different international agreements has emerged. While TRIPS and CBD deal with a patent system and country's sovereignty over genetic resources, UPOV's objective is free access to genetic resources for research and development (Virchow, 1999a).

However, the policy is yet too broad and less detailed in indicating the implementation of different crop diversity conservation and utilization strategies. On-farm conservation is yet lacking strong policy and institutional support. For effective implementation, the role of the different stake-holders (farmers, MoA, IBCR, EARO, EPA, breeders, seed enterprises, NGO conservators, and international community) in conserving and utilizing crop diversity will have to be worked out.

The conservation and development benefits of newly emerging technologies have to be assessed and evaluated. For instance, agricultural biotechnology has both potential utilities (in the areas of agricultural production, food security, poverty reduction and even conservation of crop diversity) and potential risks (health and environmental). The country will have to develop a biotechnology policy to guide its future action plans in this emerging and controversial scientific advance so that it can reap its benefits.

Clearly, addressing each of the above constraints requires innovation in a diversity of institutions. To contribute to this challenge, this thesis is mainly concerned with the mechanisms of enhancing the role of farmers to maintain crop diversity and the role of crop diversity to their livelihoods.

4.6. Selected economic theories relevant for on-farm conservation research

In this part of the thesis, the relevant economic theories to study the economics of on-farm conservation of crop diversity are presented. These theories will be applied in the forthcoming empirical chapters.

4.6.1. The economics of on-farm conservation of crop diversity

The economics of biodiversity is concerned with how biodiversity contributes to economic activity and how economic activities, in turn, impact on the status and integrity of biodiversity (IUCN, 1999). What does the economics of on-farm conservation of crop diversity stand for in the thesis? It refers to the study of farmers' incentives/dis-incentives, their attribute preferences, the opportunity costs of planting local varieties, and the implications of all these results for on-farm conservation.

The conservation and sustainable use of crop diversity is a matter of public policy, international law, politics, ecology and economics. In large measure, biodiversity conservation has often been treated as a biological issue – a question for biologists to debate, offer counsel about, and act upon (Perlan and Adelson, 1997). But economics plays an important role both as an integral part of other concerned disciplines as well as independently.

Economics can have a role in the following key areas of crop diversity research and policy: measuring costs and benefits of maintaining local varieties both to the farmers and the nation at large, understanding the socioeconomic factors affecting farmers' variety use decision, understanding farmers' variety attribute preferences, designing the right incentive mechanisms required to achieve both crop diversity conservation and economic development, computing and analyzing opportunity costs of conservation of crop diversity, and defining operational concepts to deal with the intricacies of property right issues.

4.6.2. The theories of impure public goods and joint production in agriculture

The in-between points along the pure private and pure public goods continuum are occupied by the so-called *impure public goods* (*IPGs*). Their benefits are partly rival and/or partly excludable. For the purpose at hand, indigenous sorghum or coffee farming is doing two things in one: producing crops as private consumption goods and contributing to the diversity of the respective crops as *IPGs*. Essentially, farmers' activities are resulting *joint products i.e.* coffee cherries as private consumption goods (either to the market or own use) and the genetic information embedded therein as *IPGs*. The theory of *IPGs* is originally developed by Cornes and Sandler (1986).

Beyond its primary function of supplying food and fibre, agriculture also provides environmental benefits (such as preservation of crop diversity). This is what is called the *multifunctional* character of agriculture (Maier and Shobavashi, 2001). *Multi-functionality* or *joint production* in agriculture is concerned with the production relationships underlying the multiple outputs of agriculture, and the externality and public good aspects of the non-commodity outputs of agriculture. These outputs can be positive or negative, intended or unintended, complementary or conflicting, and reinforcing or offsetting.

Accordingly, all farmers in a given region or nation are cumulatively contributing to the regional or national level crop diversity. The production of the *joint product* by each farm household is a function of the respective utility functions reflecting their *revealed preferences* from the available choice set. Of course, crop diversity is not part of farmers' direct utility function. It is a derived demand emanating from farmers' ambitions to satisfy diverse household needs. Benefits from multiple variety attributes induce them to maintain crop diversity on-farm.

While the consumers of pure private goods have market price to influence the size of goods produced, the consumers of crop diversity as *IPGs* (private breeders and pharmaceutical companies) do not have the incentive to optimize the production of crop diversity. In the absence of property rights and with no payment for *de facto* conservation, private companies will have the incentive to free ride on the sub-optimal level of crop diversity produced.

The *IPGs* model has been used in crop diversity research, among others, by Heisey *et.al* (1997) and Smale *et.al* (2001). Accordingly, farmers derive utility not only from a pure private good but also from their own contribution to the public good. For instance, when there is a trade-off between yield potential and insurance value of crop diversity, the farmer looking for better yield has an incentive to under-produce crop diversity (Heisey *et.al*, 1997).

If all outputs were private goods for which well-functioning markets exist, private transactions would ensure that supply and demand are balanced and resources are used efficiently. However, most of the multiple outputs of agriculture (in this case crop diversity) are externalities or public goods. There are no market signals for farmers to know how much of these outputs to produce. The outputs will still be generated, as they are supplied jointly with agricultural commodities and there is no way to ensure their optimality (Maier and Shobayashi, 2001). This is mainly due to the absence of clearing house to take care of possible extinction and redundancy which, in turn, calls for policy intervention to improve the outcome.

The second lesson from the theories of IPGs and joint production in agriculture is that farmers do not need external incentives or subsidies for every components of crop diversity because they produce some level of diversity without any external incentive. But the questions that get 'NO' answers from the *IPGs* theory are: '*Do farmers care about crop diversity as scientists, crop breeders, and governments do*?', '*Are all components of crop diversity conserved de facto by farmers?*' '*Is what farmers are doing enough?*', '*Are all farmers equally contributing to crop diversity?*' and '*Are there enforceable institutions to mitigate the impact of up-coming development interventions on crop diversity?*' Had the answers for these questions been '*Yes*', this thesis work would not have been started and studies on conservation incentives, contract design, property rights related to crop diversity, ex situ conservation, and genetic erosion would not have kept resource economists busy.

4.6.3. The characteristic model and its role in crop diversity and adoption research

According to the *characteristic model*, products are differentiated based on their attributes (Lancaster, 1966; Ratchford, 1975; Ladd and Suvannunt, 1976) and they are wanted because of the utilities the attributes provide. Accordingly, utilities are derived not from goods but the characteristics embedded in the goods. To the farm household, there are two major characteristics to be reconciled *i.e.* phenotypic characteristics of varieties and characteristics of farmers themselves. Farmers with similar household characteristics will have similar variety attribute preferences. The farmer is expected to be selecting the combination of varieties that provide the combination of total product characteristics to maximize utility. The *characteristic model* allows for both positive and negative characteristics to have their respective effect on the marginal utility (Hendler, 1975). This gives a chance to understand the trade-offs involved between a good attribute in yield and bad attribute in market price.

Recently, the *characteristic model* is attracting renewed interest not only in micro-economic diversity theory (for instance, Nehring and Puppe, 2002) but also in applications concerning farmers' variety preferences and land allocation decisions (for instance, Smale *et.al*, 1998; 2001). The adoption literature is being motivated by farmers' preferences for technology characteristics (for instance, Adesina and Zinnah, 1993; Adesina and Baidu-Forson, 1995). The notions of the *characteristic model* are also used in the valuation of variety attributes (for instance, Espinosa and Goodwin, 1991).

The variety attributes and utility are, however, indirectly linked through the production process, what Lancaster calls *activity vector*. Noting that what farmers are looking for at the end of the day is variety attribute, the demand for varieties can be considered as a *derived demand* revealed from farmers' preferences for variety attributes. As per the prediction of the *characteristic model*[6], all farmers are not valuing all the attributes equally due to their heterogeneity in terms of concerns, endowments and constraints. The utility

[6]The main conclusion of the *characteristic model* in the context of crop diversity is that attributes are more important than varieties (Smale *et.al*, 2001). Of course, the varieties are as good as their desirable attributes and they are as bad as their undesirable attributes.

of variety attributes varies across farm households depending on concerns, endowments and constraints. This motivates one of the objectives of Chapter 6 to find correspondence (if any) between farmers' characteristics and their variety attribute preferences. Subject to the exogenous factors affecting utility, farmers' demand for a variety is determined by the attributes it embeds, the importance of the attributes in addressing the goals of the farm household, and the importance to the farmer of the supply of attributes in the village (Smale and Bellon, 1999).

Farmers, accordingly, are maximizing their household utility by combining the preferred variety attributes. A variety that embeds an attribute that contributes more to their utility gets more share of their resources in production and *vice versa i.e.* farmers' decision on land allocation to a certain variety is the reflection of the marginal contribution of the attribute(s) that the variety contains to the utility of the household. The survey results in Chapter 3 show that farmers use variety attributes in deciding on variety maintenance, replacement and use. Most importantly, variety attributes are important to the extent that farmers are naming the varieties based on the attributes they have attached to the varieties during their life long farming experience.

From the notions of the *characteristic model*, variety attributes (T_i's) lead to the variety use decision. This, in turn, leads to production from each crops and varieties (Y_i's). As a result, farmers' utility function can be represented as a function of variety attributes preferred (T_i) *i.e.*

$V(T_i / \varpi_i)$ Direct utility from variety attributes preferred (4.2)

V is affected by a set of exogenous factors (ω_i) including household contextual characteristics. As the household is consuming and producing its own output, both attributes affecting consumption and production should be part of the utility maximization.

Let X^1 to X^n denote the vectors of household consumption goods (own produced and purchased) considered in conventional utility functions and $V_{FV_1^1}$ to $V_{FV_n^n}$ denote the vectors of local crop variety attributes consumed by the respective households. Then the utilities of the 'n' respective households can be written as:

$U^1(X^1 + V_{FV_1^1})$ to $U^n(X^n + V_{FV_n^n})$ Utility of 'n' households (4.3)

Maximizing utility subject to the budget and technology constraints yields the 'optimum' level of land for the respective varieties, 'optimum' level of production from each crop variety and the attributes embedded in the respective varieties. Number of FVs maintained on-farm, area share of each variety, types of variety attributes embedded in FVs, and farmers' variety attribute preferences are all outcomes of the constrained utility maximization process of production. In essence, mapping is being made from household characteristics to variety attributes, resulting in demand for FVs, variety attribute preferences and on-farm crop diversity.

4.6.4. The random utility theory and its role in crop diversity research

The random utility model (Thurstone, 1927; McFadden, 1974) assumes that a farmer has a set of options to choose among which he/she can discriminate. A farmer plants a given set of varieties and the utility he/she derives by planting that particular set is greater than the utility that he/she would have from other combinations. If a farm household has three mutually exclusive options, the observed choice among the three reveals the choice that provides the greatest utility to the household.

The probability that choice T is selected within the choice set S is given by the maximum utility among all the possible combinations (U_{c1}, U_{c2}, …., U_{cn}) or symbolically:

$$P_S^i(T) = P\left(U_T^i = \max(U_{c1}, U_{c2}, \ldots, U_{cn})\right) \quad \text{Choice of attributes (4.4)}$$

The outcome of the choice is random as it is not possible to predict with full certainty the choice that a randomly selected farmer will make due to unobserved variety attributes and unobserved farmer attributes. The choices depend on both observable and unobservable characteristics of the farmer and the attributes of varieties available to him/her.

The utility that farmer '*i*' is associating with any alternative 'a' is given by two components:

$$U_a^i = \mu_a^i + \varepsilon_a^i \quad \text{Random utility (4.5)}$$

where the two terms in the right are the deterministic and the stochastic elements of the random utility function.

Suppose the farmer has two alternatives to choose with a choice set given by C = [FVs, IVs]. The probability that he will choose the FVs can be given by:

$$
\begin{aligned}
P_{\{IVs,FVs\}}(FVs) &= P\left[U^{i}_{FVs} \geq U^{i}_{IVs}\right] \\
&= P\left[\mu^{i}_{FVs} + \varepsilon^{i}_{FVs} \geq \mu^{i}_{IVs} + \varepsilon^{i}_{IVs}\right] \\
&= P\left[\mu^{i}_{FVs} - \mu^{i}_{IVs} \geq \varepsilon^{i}_{IVs} - \varepsilon^{i}_{FVs}\right]
\end{aligned}
\qquad (4.6)
$$

To derive the farmer's preference, the usual assumptions on the random term (zero mean, constant variance and a specified functional form) should be imposed. Regarding the deterministic term, the utility levels are functions of the capacity of attributes of the varieties to fit into farmers' characteristics. Thus, variety attributes, farmers' characteristics, and the decision rules are the major ingredients of the random utility model.

4.6.5. The theory of incentives and conservation of crop diversity

It is well documented and understood that individuals respond to economic incentives and expectations (Furubotn and Pejovich, 1974). The principle of incentives is that farmers themselves must perceive an advantage in continuing to grow local varieties and their participation in conservation must be self-sustaining. In short, their expectations must be met.

The term 'incentives' in this thesis is used to represent two inter-related points: contextual factors derived from within the household and motivating farmers to make pro-diversity decisions and secondly the required incentives from policy to close the gap between *de facto* conservation and the 'optimum or required' level of conservation. For the sake of convenience, the former are called '*in-built incentives*' and the later are coined as '*policy incentives*'. Addressing *perverse incentives* is as important as designing policy incentives. The main argument of the thesis throughout is that understanding the in-built incentives, preferences and opportunity costs is a pre-requisite for the design of successful policy incentives.

Economists often tend to focus on so-called economic incentives. Incentive measures are, however, more than economic instruments involving cultural norms and traditions. If an incentive measure is to work, it must make sense economically, socially, culturally, and legally. The effect of various in-built incentives on farmers' demand for local varieties, variety count, attribute count and variety attribute preferences will be empirically examined in Chapters 5 and 6.

4.7. Summary and concluding remarks

This chapter has outlined the current conservation paradigms, issues that have to be addressed in crop diversity conservation, farm level factors potentially leading to crop diversity loss, national level factors complicating crop diversity conservation in LDCs, and the utility of economics and economic theories in crop diversity conservation, research and policy. The outcome of this assessment is that there are still inherent challenges forcing society to go beyond economic calculations. Moreover, while the household level factors are less worrisome at least in the short-run, the national level factors are too complex.

At the end, this chapter has presented economic theories relevant for research on the economics of crop diversity. The subsequent chapters will apply the theories in analyzing farmers' in-built incentives, variety attribute preferences, demand for local varieties, and opportunity costs of maintaining local varieties on-farm.

5. Economic analysis of farmers' incentives to diversify on local sorghum varieties

5.1. General background

For those local varieties worth conserving and not maintained *de facto*, policy incentives easy to implement and designed based on knowledge about farmers are needed. Still, before designing sound policy incentives for on-farm conservation and/or removing perverse incentives, understanding why farmers are doing what they are doing (regarding crop variety management) is crucial. There is a need to understand what factors motivate them to produce crop diversity *de facto* and how farmers' resource endowments, constraints, and working environment result in crop diversity on-farm.

Many studies on crop diversity acknowledge that farmers are pioneers in conserving local varieties (Worede, 1988; Wood and Lenne, 1997; Teshome *et.al*, 1999). In such studies, all farmers are equally praised for their contribution in maintaining crop diversity. This chapter will give explicit account of farmers' heterogeneity (in terms of resource endowment, sensitivity to risk, concerns, objectives and constraints) and examine their in-built incentives and dis-incentives for managing a portfolio of local varieties. The research question to be addressed is, therefore, 'What factors induce on-farm *de facto* conservation of sorghum diversity and which ones do have the opposite effect?'

In this chapter, the thesis shall:

- analyze the mechanisms by which farmers' 'survival first' motives promote crop diversity;
- examine the farm household contextual factors inducing farmers to maintain multiple local varieties of sorghum; and
- examine factors that detach the link between *de facto* conservation and farmers' survival strategies.

In the theoretical modeling, *de facto* conservation shall be considered as a positive externality of farmers' resource allocation decisions. Of course, for the purpose at hand, it is neither interesting nor possible to precisely determine crop genetic diversity on-farm.

The remaining part of this chapter is structured as follows. Section 5.2 presents the role of studying in-built incentives for on-farm conservation. Section 5.3 details the theory to be followed by the econometric model in Section 5.4. The variables are described in Section 5.5 and regression results are presented and discussed in Section 5.6. Concluding remarks follow in the last section.

5.2. Why in-built incentives matter?

Both improving agricultural productivity and sustainable resource management are recognized major policy and research issues to address rural development in Ethiopia (Omiti *et.al*, 2000). However, most of the research so far concentrates on farmers' preferences for new varieties and identifies *ex ante* or *ex post* constraints of their adoption. This emphasis, however, does not give a complete picture of farmers' variety choice.

In the process of *de facto* conservation, there could be local varieties not of interest for any farmer (resulting in possible extinction) and there could be local varieties of interest for thousands of farmers (resulting in redundancy). As Maier and Shobayashi (2001) put it succulently 'it would be sheer coincidence if the level, composition and quality of biodiversity corresponded to those demanded by society. Thus, farmers' contribution without policy incentives will not necessarily be adequate. This calls for the need to change the *status quo*.

For those local varieties that deserve conservation, designing incentives or removing perverse incentives (McNeely, 1988) are needed. This, among other things, entails identifying representative farmers. Before any intervention, decision-makers have to be informed on farmers' in-built incentives to undertake poly-variety. Characterizing farmers and studying their 'survival first' motives for undertaking on-farm conservation is a crucial step to identify farmers to be targeted and influence or predict their variety use behavior. This enquiry is also important to design contracts with farmers targeted for on-farm conservation (Edilegnaw *et.al*, 2003).

To sum up, understanding farmers' incentives to manage multiple local varieties is important, among other things, to identify farmers for on-farm conservation, to harmonize conservation endeavors with their objectives, to locate where government can do on-farm conservation by itself and where it can better delegate farmers, and to design contextual policy incentives to be put in place.

5.3. The theory: deriving diversity from household utility

Why do farmers grow multiple local varieties? For a farmer to opt for a single variety, the following ideal conditions have to be in place. A single variety should satisfy all the farm household needs. There should be a well functioning market to exchange the different varieties and farmers should be indifferent between using own produced and bought crop varieties (perfect substitutability). Their input endowments should be equally suitable for all varieties (absence of *asset specificity*). Had there been constant returns to scale for every variety and had farmers been able to satisfy their production and consumption needs with a single variety, specialization would have been the rule and diversity on each farm would have been zero. Last but not least, farmers should have a single profit maximization objective. These conditions do not hold in almost all the cases and hence most farmers are managing a portfolio of varieties to address their household needs[1].

So far, theories of farmers' behavior under risk (e.g. Dercon, 1996; Rothenzweig and Binswanger, 1993), access to markets (e.g. de Janvry *et.al*, 1991), and environmental constraints (e.g. Bellon and Taylor, 1993) have been forwarded to explain the resource allocation decisions of farm households. The implications of these models for in situ conservation have, among others, been studied by Meng (1997), Smale *et.al* (2001), and van Dusen (2000). As far as designing incentive mechanisms for on-farm conservation of crop diversity is concerned, it is important to note that a change in any single economic factor is unlikely to cause farmers to cease or start growing local varieties (Meng, 1997; Smale *et.al*, 1994). Hence, any model to explain *de facto* conservation should be as comprehensive as possible. Because the alternative explanations are not mutually exclusive, failure to estimate a comprehensive model overstates the significance of an individual explanation. Thus, no single theoretical argument fully explains farmers' land allocation decisions (Smale *et. al*, 1994). In econometric terms, partial analysis will bring the problem of omitted variables resulting in biased estimates.

Economic models for analyzing crop choice and diversity are derived from the partial adoption literature (Feder *et. al*, 1985 and Feder and Umali, 1993) and agricultural household models (Singh *et. al*, 1986). If the cultivator's technology choices are the result of profit maximizing behavior, then varietal choice will depend on the determinants of profit (Feder and Umali, 1993). In Chapter 3, it was pointed out that farmers are growing multiple

---

[1] All these conditions are not mutually exclusive.

varieties for reasons far more complex than just maximization of profit. Thus, a micro-economic model based on maximization of profits *per se* may not explain why these farmers grow multiple local varieties. Consequently, a utility-based model that allows trade-offs among different economic and non-economic sub-objectives is used.

Economic theory suggests different factors that affect farmers' crop and variety choice. Reviewing the available literature reveals that the possible factors are access to markets, income diversification motives, heterogeneity of farmers' resources (mainly land), resource endowment (education, labor, and wealth), multiplicity of farmers' household needs (livestock ownership and risk), and the impossibility to address all these with a single crop variety.

The effect of household resource endowment (like labor) on farmers' incentives to diversify depends on the labor intensity of growing different crops and set of varieties. Farmers' experience suggests that growing IVs and/or multiple varieties (local or improved) is labor intensive. The IVs are labor demanding because they demand intensive application of cultural practices. The multiple varieties are labor demanding because they require different treatments in agricultural practices at different points in time. Therefore, *ceteris paribus*, more labor endowment will translate to using multiple varieties and/or taking IVs. This makes it difficult to predict its impact on the production of crop diversity.

The theoretical model sketched below is meant to incorporate each of these variables step by step. The analytical framework presented here derives on-farm crop diversity as the result of farmers' utility maximization constrained by survival motives. The model formulation follows van Dusen (2000) who derives the diversity function integrating missing markets and risk. The main difference lies in our handling of risk using the safety first notion, and not just wealth which was the risk proxy in previous studies by van Dusen (2000) and Meng (1997), for instance. Assuming a well behaved utility function, the farmers' utility maximization problem can be set as:

$$Max{:}\ U = U(G_f, G_p / \varpi_0) \qquad \text{Utility function (5.1)}$$

where G_f and G_p refer to consumption of aggregate on-farm produced and aggregate purchased goods, respectively[2]. ϖ_0 refers to the vector of household level exogenous factors including input endowments and heterogeneity (land and labor), human capital (farmers' age and education), asset endowment (livestock, land holding and wealth), and access factors (extension and credit)[3] conditioning utility.

The utility is subject to the following constraints:

$$p_g(G - G_f) - c(G;\varpi_0) + \ddot{Y} \leq Y \qquad \text{Full income constraint (5.2)}$$

where G, p_g, $c(.)$ and $\ddot{Y}$ are the aggregate produced good on-farm, price of the good, cost of production for the good and income from sources outside agriculture, respectively. The full income constraint encompasses cash, labor and livestock feed constraints (if the latter two have to be purchased).

$$CR + C_{own} = \sum_{i=1}^{n} C_i * E_i + \sum_{i=1}^{k} CP_i * G_{pi} + R \leq CASH \qquad \text{Cash constraint (5.3)}$$

where CR, C_{own}, C_i, CP_i, R, and $CASH$ denote cash borrowed from formal or informal sources, cash from own sources, cash spent on the ith farm enterprise (E_i), cash spent on the ith consumption good (G_{pi}) per unit, cash reserved for unforeseen expenses, and the total cash available at the disposal of the household, respectively.

$$L_F + L_{leisure} + L_{NF} + L_{Hired} \leq LABOR \qquad \text{Labor constraint (5.4)}$$

where L_F, $L_{leisure}$, L_{NF}, L_{Hired}, and $LABOR$ refer to labor used on the farm, labor spent for leisure, labor spent for non-farm economic activities, hired labor, and household labor endowment, respectively.

$$LA_{own} + LA_{rentin} - LA_{rentout} = \sum_{i=1}^{n} d_i E_i + LA_{fallow} \leq LAND \qquad \text{Land constraint (5.5)}$$

---

[2]Quantities of goods produced are aggregated here for simplifying the exposition. Multiple output production can also be derived similarly (Singh and Subramanian, 1986).

[3]Note that market access is not yet part of the model.

where LA_{own}, LA_{rentin}, $LA_{rentout}$, LA_{fallow} and *LAND* are own land operated, rent-in land, rent-out land, land left fallow, and total land endowment. *d*, *E* and *n*, are the size of land allocated for the ith enterprise, crop or livestock enterprise, and number of enterprises that the household deals with, respectively.

The other most important constraint that the household faces is the production technology constraint *i.e.*

$$P(G_f, G_m) = 0 \qquad \text{Production technology constraint (5.6)}$$

where G_m refers to aggregate on-farm produced commodity sold in the market[4]. The production technology is conditional on choices (varieties and crop enterprises). Assuming perfect markets and risk free environment, the first order conditions of this problem lead, among other things, to the following optimum level of production of the market good (G_m^*) and on-farm consumed good (G_f^*):

$$G = G(G_m^*, G_f^*; \varpi_0) \qquad \text{Optimum production (5.7)}$$

The household chooses a set of varieties, crop enterprises and inputs to get the optimum level of production from each crop enterprise (G_f^*, G_m^*). While trying to meet their household objectives, farmers not only produce and consume crops but also maintain diverse set of local crop varieties year after year as a positive externality of their land use decisions.

Utility maximization generates not only optimum level of production but also derived demand for crop diversity (D_c) or number of varieties and crops planted subject to the factors motivating farmers to diversify *i.e.*

$$D_c = \left[\sum_{i=1}^{n} v_i / \sum_{i=1}^{m} CE_i ; \varpi_0 \right] \qquad \text{Derived demand for diversification (5.8)}$$

where v_i's index local varieties and CE_i's index crop enterprises selected from the available choice set, respectively, so that v_i and CE_i take 1 if the household chooses the variety or the crop enterprise, and zero otherwise, respectively. As we do not observe D_c directly in the household's decision process, van Dusen (2000) has called it *latent diversity*. It is the outcome of

4 Labor, land and capital constraints are part of the production technology constraint. Capital is a fixed input and it is left out for simplification.

farmers' revealed preferences to meet their household concerns. If the ideal conditions outlined at the beginning of this section are in place, the outcome of a risk free environment will be specialization and no crop diversity on-farm[5].

To relate crop diversity index estimation with farmers' derivation of household utility, let us consider a farmer planting only sorghum and getting income only from this crop. If the farm household needs are met from planting '*n*' varieties, then all the '*n*' varieties are making a certain contribution to satisfy household concerns. The contribution of each variety to household concerns is an increasing function of the share of the resources it takes. If a farmer decides to have a single variety, he/she is investing all his/her resources (meant for that crop) to that variety. Consequently, he/she is satisfying all his/her household needs from that single variety. On the other hand, if he/she opts for multiple varieties, then all the varieties are sharing the resources and cumulatively contributing to his/her livelihoods. Hence, the proportion of land allocated to a variety, which is used to compute ShaNNoN's diversity index (Sub-section 4.3.5), is indicating the relative contribution of the variety to farmers' utility derived from sorghum.

Given farmers' working environment, the genetic composition of crop diversity on- farm (D_{cf}) is a function of farmers' on-farm management practices (M_p) and natural factors affecting the ecological interactions (N_f) subject to ϖ_0 *i.e.*

$$D_{cf} = f(D_c; M_p, N_f / \varpi_0) \qquad \text{Genetic diversity on-farm (5.9)}$$

So far, access to markets is not part of the model. Market access and lower transaction costs nullify the need for farmers to be self-sufficient (de Janvry *et. al*, 1991). When farmers have access to markets and roads, many of the services provided by crop diversity can be substituted by alternatives that can be purchased in the market and as a result the production of crop diversity will shrink (Bellon, 1996). This may lead to uniformity in crop variety (ies) on-farm.

Missing markets bring additional input and output constraints by forcing farmers to be self-sufficient *i.e.* balance the quantity of non-marketed goods (NM_g) to household demand (HH_{dd}), labor demand to own labor supply, cash

[5]Crop diversity may or may not occur at the regional or national level depending on the production of crop diversity across farm households.

demand to own cash, and land use to own land. Thus, utility maximization is further constrained by:

$NM_g = HH_{dd}$ Output constraint (5.10)

As a result of this constraint, the production technology constraint above also includes NM_g. The other three input constraints called by missing markets include:

$C_{own} = \sum_{i=1}^{n} C_i * E_i + \sum_{i=1}^{n} C_i * G_{pi} + R \leq CASH$ Cash constraint (5.11)

$L_F + L_{leisure} + L_{NF} \leq L_{HH}$ Labor constraint (5.12)

$\sum_{i=1}^{n} d_i E_i + L_{fallow} \leq L_{own}$ Land constraint (5.13)

The optimum level of production of the traded and non-traded goods with incomplete markets will then be:

$G = G(G_f^*, G_m^*, NM_g^*; \varpi_1)$ Optimum production with missing markets (5.14)

where ϖ_1 is ϖ_0 plus input (equations 5.11 to 5.13) and output (equation 5.10) market related constraints conditioning utility.

Replacing ϖ_0 with ϖ_1, the crop diversity outcome represented by equations (5.8) and (5.9) will then follow. Complete absence of a market is rather an exception and usually farm households are faced with relative degree of detachment from the market. Taking this into account, the variable considered to proxy market access will be the walking distance from the farm stead to the nearest market.

The next most important variable worth considering is risk. Farmers' risk aversion behavior is governed by the extent to which risk affects their livelihood. While risk is forcing farmers to take cautionary measures, their reaction to it mainly depends on their sensitivity to potential shocks which, in turn, is highly dependent on their wealth status and farmers' expected risk free income. The higher farmers' vulnerability to potential income shocks, the more risk averse they will be. While less-vulnerable farmers can make consumption smooth, vulnerable farmers are highly sensitive to potential farm income variability. Moreover, the marginal pain from the potential loss is more harmful for the resource poor farmers. How do such farmers deal with such a reality?

One of the indigenous means to deal with risk associated with farming is local variety portfolio management. Managing a diverse set of local varieties of crops enables farmers tolerate potential income shocks. To study the linkage between farmers' variety diversification and risk, this chapter takes the results of previous studies that synthesize relationships among risk, consumption smoothing, crop choice, and wealth (Rosenzweig and Binswanger, 1993; Murduch, 1995; Dercon, 1996) and integrates the basic notions of Roy's safety first approach (Roy, 1952) into the model outlined above[6].

Considering risk as a production problem that involves yield variability, farmers are trying to have large portfolio of crop varieties to spread possibility of risk occurrence[7]. According to the portfolio theory, the demand for crop diversity depends on the effect of the diversity on total yield variability. If yield variability decreases as a result of crop diversity, then derived demand for crop diversity will be an increasing function of farmers' sensitivity to potential income shocks[8].

Farmers are assumed to be trading high return for stable income. According to the '*safety first*' approach, each farm household head is assumed to have a target level of income in mind. This is the minimum income necessary for survival that the household has to achieve to feed, shelter and cloth all household members. If there is higher expected net return with higher yield risk (say from adopting IVs) and/or lower expected net return with lower yield variability (say from using multiple FVs)[9], the decision of the farmer

[6]There are different reasons for the choice of this method. The theory is consistent with small scale peasant household decision behavior as farm households are maximizing not income but the probability of survival. The other desirable feature of this approach is that different choices made by farmers depend on farmers' differences in their subsistence needs, resource endowments, and perceptions of riskiness among competing activities (Parikh and Bernard, 1988).

[7]Diversification does not work when risks are covariate *i.e.* risk occurrences that could happen together like loss of sorghum and maize as a result of drought.

[8]It is important to bear in mind that income diversification for risk reason does not necessarily mean foregoing income. By the same token, income source specialization does not necessarily mean earning extra income.

[9]The intensive discussions made with farmers during the survey confirm that farmers perceive growing IVs as more risky and growing multiple local varieties as less risky.

considering survival first depends on the extent to which the household is able to fulfill basic household needs ($Basic_{req}$) from its internal endowment (wealth plus expected non-covariate income[10]) denoted as FN_{income}. This can be considered as income and wealth at the disposal of the farm household that it will have to resort to in case crop fails. Those farm households who have higher FN_{income} will be less worried about future yield variability due to crop failure. Having better FN_{income} status, they will feel more secured. Farmers' objective is thus to minimize the probability that FN_{income} falls short of $Basic_{req}$ *i.e.*

$$\text{Min } P(FN_{income} < Basic_{req}) \qquad \text{Farmers' 'survival first' motive (5.15)}$$

Accordingly, farmers will gamble with nature if $FN_{income} > Basic_{req}$ and they will take more cautionary measures if $FN_{income} < Basic_{req}$. In our case, FN_{income} is computed as the sum of the value of livestock, annual income from cht'at, and annual estimated income from non-farm income sources[11].

Given this framework, risk is proxied by the extent to which the farm household is able to satisfy $Basic_{req}$. Putting both FN_{income} and $Basic_{req}$ on per-capita basis[12],

$$Risk_{proxy} = SAFTYPC - REQPC \qquad \text{Risk derived from 'survival first' (5.16)}$$

where $SAFTYPC$ and $REQPC$ are FN_{income} and $Basic_{req}$ divided by household size equivalents, respectively. Farmers' courage and level of curiosity to take more risky ventures is better depicted as a continuum depending on the size of the surplus (positive $Risk_{proxy}$) and deficit (negative $Risk_{proxy}$). Encompassing $Risk_{proxy}$ in the constrained utility maximization results in the optimum level of production:

---

[10] Non-farm income which has a flow not correlated to farm income.

[11] Not only income but also wealth variables are used to compute FN_{income} because farmers sell their wealth when income falls short of current consumption needs. Hence, those farmers who have more livestock to sell if crops fail are, *ceteris paribus*, considered to be more risk takers. Household level data were collected on $Basic_{req}$ and components of FN_{income}. The flow (income) and stock (wealth) concepts are summed up because considering them independently in the regression has resulted in the problem of multicollinearity. The single risk proxy is computed at the expense of adding flow and stock concepts which has to be noted as a shortcoming.

[12] Consumption requirements of household members have been adjusted using the figures in Table III-1 of appendix 3.

$$G = G(G_f^*, G_m^*, NM_g^*; \varpi_2) \qquad \text{Optimum production (5.17)}$$

where ϖ_2 refers to ϖ_1 plus the 'safety first' household concerns.

In the context of the utility function, $Risk_{proxy}$ conditions utility maximization and variety choice through the 'risk free' full income constraint whereby farmers' sensitivity to the effect of potential farm income shock is measured by the extent to which full income (FI) satisfies family requirement *i.e.*

$$P_c Cht'at - c(Cht'at) + \sum_{i=1}^{m} P_{ki} K_i + \ddot{Y} = FN_{income} \qquad \text{FI conditioning risk behavior (5.18)}$$

where P_c, $c(.)$, P_{ki}, m and $\ddot{Y}$ refer to price of cht'at, cost of production for cht'at, price of the ith livestock (K_i), the number of livestock per household, and unearned income and/or non-farm income.

Derived from the theory adopted, the hypotheses to be tested in this chapter include:

- Lack of access to markets and land heterogeneity induce farmers to undertake poly-variety;
- Better access to markets for high-value cash crops and experience in using IVs threaten the diversity of the less-valuable food crop; and
- The more the deficit that farmers have to deal with to satisfy subsistent requirement (more negative $Risk_{proxy}$), the higher will be the portfolio of local varieties maintained on-farm and *vice versa*.

5.4. The econometric model

Farmers' variety management decisions involve both discrete choice on each variety (whether or not to plant a given variety?) and continuous decisions (how much land to allocate or how much of each variety or crop to plant?). Crop diversity is mainly a function of the discrete decision (van Dusen, 2000). To explain the factors affecting D_c, the process of crop diversity production shall be considered as the outcome of a set of discrete choice decisions by farmers. Suppose the total level of crop diversity in a given locality is 'D_R'. If every local crop variety contributes to 'D_R', then each farmer's discrete variety choice decision is contributing to 'D_R'. Accordingly, the number of varieties and/or crop enterprises on each farm is:

$$D_c = \left[\sum_{i=1}^{n} v_i / \sum_{i=1}^{m} CE_i ; \varpi_2 \right]$$ Derived demand for diversification (5.19)

where v_i's ($v_i = 1$ if the variety is selected and zero otherwise) stand for FVs and CE_i's ($CE_i = 1$ if the variety attribute is important to the household and zero otherwise) stand for crop enterprises. '*n*' and '*m*' are the maximum number of FVs and crop enterprises available in the village, respectively. Farmers make descrite choice decisions based on the available choice set.

Thus, there are multiple variety use decisions (by millions of farmers) resulting in multiple varieties on-farm (D_c) and cumulatively contributing to 'D_R'. The summation of a series of discrete choices can be approximated using a Poisson regression for a count of the total number of varieties produced (Pudney, 1989; Hellerstein and Mendelsohn, 1993).

Repetition of a series of binomial choices by thousands of farmers asymptotically converges to a Poisson distribution as the total number of local varieties counted on all farmers' fields and the proportion of sorghum varieties on a given farm that contribute to crop diversity gets large. This well known relationship between the Poisson and binomial distribution (Pudney, 1989) justifies our choice of the Poisson regression model (Cameron and Trivedi, 1998) to explain on-farm count diversity. Using the Poisson model is justified not only because it is statistically consistent but also the number of FVs maintained on-farm (diversity index used in this chapter) is integer count. Count diversity index is computed as the number of FVs identified minus one (Taillie and Patil, 1982).

The Poisson regression model assumes that y_i (variety count) given x_i (factors affecting variety count) is Poisson distributed with density:

$$f(y_i / x_i) = \frac{e^{-\lambda_i} \lambda_i^{y_i}}{y_i!}, \quad y_i = 0,1,2$$ Poisson distribution (5.20)

and mean parameter,

$E[y_i|x_i] = \lambda_i = exp(x_i'\beta) = exp(\beta_0)exp(\beta_1 x_{1i})exp(\beta_2 x_{2i}) \ldots exp(\beta_k x_{ki})$.

Having independent observations, the model to be estimated is:

$$y_i = e^{x_i'\beta} + \varepsilon_i = e^{(\beta_0 + \beta_1 x_{1i} + \beta_2 x_{2i} + \ldots + \beta_k x_{ki})} + \varepsilon_i$$
$$\Rightarrow \ln y_i = \beta_0 + \beta_1 x_{1i} + \beta_2 x_{2i} \ldots + \beta_k x_{ki} + \varepsilon_i$$ Poisson regression model (5.21)

where the x_i's are the variables (described in Table 5-1) affecting on-farm sorghum diversity (COUNT). Differentiating the second equation of 5.21 with respect to x_{ji} yields,

$$\beta_j = \frac{\partial E[y_i \mid x_i]}{\partial x_{ji}} \frac{1}{E[y_i \mid x_i]} = \frac{\partial \ln E(y_i \mid x_i)}{\partial x_{ji}}.$$

The coefficients of the Poisson model can, therefore, be interpreted as the proportionate change in the conditional mean if the jth regressor changes by one unit *i.e.* semi-elasticity. Differentiating the first equation in 5.21, the marginal effects in the Poisson model are computed as:

$$\frac{\partial E[y_i \mid x_i]}{\partial x_{ji}} = \exp(x_i'\beta) \times \beta_j \qquad \text{Marginal effects in Poisson (5.22)}$$

The Poisson model is, however, restrictive as it imposes the restriction that the variance of the data is equal to the conditional mean (*i.e.* $var(y_i/x_i, \beta) = E(y_i/x_i, \beta)$. To check this in the data set, regression based tests for over or under-dispersion suggested by Cameron and Trivedi (1990) are followed *i.e.* given that $Var(y_i/x_i, \beta) = \alpha \times \exp(x_i'\beta)$, if $\alpha = 1$, the data are equi-dispersed and if $\alpha > 1$ or $\alpha < 1$, the data are over-dispersed or under-dispersed, respectively. All the test results consistently show under-dispersion in our count data. Consequently, the estimates from the poisson regression will be consistent but inefficient (Gourieroux *et.al*, 1984). Moreover, the standard errors will be biased downward (Cameron and Trivedi, 1986) resulting in spurious acceptance of a non-significant coefficient to be significant. For this reason, the standard Poisson regression model has not been appropriate for our purpose.

The easiest way to deal with this problem is to use the Poisson model with the robust[13] (sandwich) covariance matrix because the Poisson model stays consistent under the violation of the equi-dispersion assumption (Winkelman, 1995). The robust procedure satisfactorily computes covariance matrix for correct inference. Hence, to deal with the violation of the equi-dispersion assumption of the Poisson regression, the robust method

[13]The robust estimation is the model estimation that accounts for heteroscedasticity and autocorrelation by transforming the dependent variable, the regressors, and the error term. This estimation enables us to make statistical inferences even if the data are not independently and identically distributed (StatCorp, 2001).

of estimation has been used to compute the Huber/White robust standard errors reported in Table 5-2 below.

To account for evenness and dominance and to compare with the Poisson regression results of variety count, other diversity indices have been computed and the results of a two-limit Tobit model have been presented in Tables IV-3 and IV-4 of Appendix IV-A. For long-term on-farm conservation decisions, the count variety index is more appropriate as it reflects farmers' life long decisions to maintain their varieties.

5.5. Describing the variables used in the regression

This section will define the variables used in the regression and give descriptive statistics of the variables employed. The data generation process and sorghum farmers' variety choice behavior have been described in detail in Chapter 3.

The diversity of purposes of growing a crop has a direct impact on the number of varieties to be used on-farm as the survival of the household members will heavily depend on the crop. Farmers have reported six major purposes (food, cash, animal feed, fuel wood, house construction and fence construction) of growing sorghum in the study area. Among the sampled farm households, 28.3, 25.3, 18.7, 20.7, and 7.1 percent of them have been producing sorghum for 1, 2, 3, 4, and 5 purposes, respectively. The average number of FVs of sorghum grown on-farm is 1.84. Among the sampled farm households, 74 of them had reported to have a single local variety; the other 84 had 2 varieties; 33 of them had 3 local varieties; and the last 5 had 4 varieties. For the estimation purpose, the last ones are censored to 3 since they are very few.

As to the village differences in variety count index, in Kero Deda (where there is no much experience with IVs), 5.9, 50, 35.3 and 8.8 percent of the farmers have reported to have been planting 1, 2, 3 and 4 local varieties, respectively. In Asseliso, where there is a high rate of distribution of IVs, 48.6, 42.9, and 8.6 percent of them have reported 1, 2, and 3 FVs, respectively. These descriptive results tentatively show the possible 'replacement' of FVs by IVs over the long-term. Table 5-1 below defines the variables used in the regression, provides the expected signs based on the theoretical predictions along with the descriptive statistics for the respective variables.

Table 5-1.Variable definitions and expected signs

Variable name	Description	Mean	SD	Expected sign
Count	The number of FVs on-farm less one	0.81	0.74	DV
Explanatory variables				
Age	Age of the household head (Years)	41.4	11.97	+
Totlabor	Total household labor endowment	3.53	1.93	+/-
Lancrops	Quantity of land allocated for all crops (Hectares)	1.51	0.96	+
Nopurpos	Number of purposes for which sorghum is used	2.54	1.31	+
Nonffarm	Income source outside agriculture (Dummy)	0.41	0.49	-
Educate	Education level of the household head	1.37	2.08	-
Impexep	Experience in growing IVs (years)	2.05	1.89	-
Credit	Participation in credit (Dummy)	0.35	0.48	-
Plots	Number of plots operated by the household	1.75	0.83	+
Access	The average of walking time needed to reach the extension agent, the nearest weather road, and market (average sum of the three in minutes)	50.24	24.6	+
$Risk_{proxy}$	Income gap per capita that each household fails to satisfy REQPC (if negative) and the *vice verse* (if positive) (Birr)	121.8	399.2	–
Cht'at	0 – no cht'at at all; 1 – only for own consumption; 2 – also for village sales; and 3 – also for sales in the cities	1.297	1.18	-

Notes: DV – Dependent variable.

Computation of 'Totlabor' excludes household's dependents (children below 9 and inactive household members because of age or permanent sickness). Household members aged 16-59 get a value 1. Children between ages 9-15 get an equivalence of 40% and household members above the age of 60 get 60% equivalence. Children going to school are assumed to spend 40% of their time to satisfy household labor demand. These all conversions are derived from previous literature and they are adapted based on the discussions made with the key informants during the survey.

The average size of the land reported by farmers to have been operated is 1.3 hectares while the average sum of land allocated for the different crops is 1.5. This difference could be either a reporting error or due to renting in and out of land or land left fallow. For the empirical analysis, the sum of land allocated for different crops is employed since it minimizes the reporting error.

Only 41% of the sampled households have reported to have earned non-farm income which is rather low. The average grade of schooling for the heads of the sampled households is 1.37 with a maximum of 11 and minimum of zero. Farmers' experience with IVs ranges from 0 to 10 with a mean of 2 years. The data suggest that 35.4 percent of the households have been involved in formal or informal credit. Farmers need, on average, about 50 minutes walk to reach the nearest market, dry weather road and, the extension agent. This figure ranges from 5 minutes to 2.5 hours.

The simple spearman correlation coefficient between *count* and $Risk_{proxy}$ is -0.25 (significant at 1%) which satisfies *a priori* expectations. The mean $Risk_{proxy}$ is positive implying that on average all households can more than satisfy $Basic_{req}$ if incomes are equally re-distributed. However, the variation is very high (ranging from - 1098 to + 1090) mainly due to the unequal distribution of income from an important cash crop in the region called cht'at. Farmers who produce and sell cht'at are typically rich which yields a positive over-all average $Risk_{proxy}$. As to the importance of cht'at to the sampled households, 41.3% of the households do not have cht'at at all; 15.7% of them produce for own consumption; 23% of them sell in the villages; and 20 percent sell in the cities[14].

5.6. Estimation results and related discussions

The variables Belinarba (24), Kerodeda (34), Asselliso (36), Ejeaneni (34), Chachole (36), and Gurbo (15) are all the village dummy variables meant to capture any village differences not accounted by the other variables[15]. The

---

[14]This variable is made categorical to reduce its correlation with $Risk_{proxy}$ and because of a clear-cut difference among the categories created. Moreover, the category better reflects their respective behavior.

[15]The numbers in brackets are the number of households sampled in each village.

village Umerkule (19) is left as a reference. The purpose of estimating both Poisson and Ordered Probit (Appendix IV-A: Tables IV-1 and IV-2) models is to check the robustness of the results. The results are by and large similar. Table 5-2 reports the parameter estimates of the Poisson model.

Table 5-2. Poisson regression results

Variable	Coefficient	Marginal effects: Dy/Dx
Age	0.0072* (1.72)	0.005
Cht'at	-0.174*** (-3.08)	-0.110
Lancrops	0.167*** (3.50)	0.105
Impexep	-0.186*** (-4.55)	0.118
Nopurpos	0.318*** (6.80)	0.201
Totlabor	0.021 (0.95)	0.013
Access	0.0026 (1.47)	0.002
Educate	0.0382* (1.69)	0.024
Plots	0.0908 (1.44)	0.057
$Risk_{proxy}$	-0.0003** (-2.24)	-0.0002
Belinarba	0.1653 (0.98)	0.111
Kerodeda	0.363** (2.20)	0.259
Asseliso	0.537*** (2.70)	0.408
Ejeaneni	0.612*** (3.19)	0.479
Chachole	-0.086 (-0.33)	-0.053
Gurbo	-0.6059 (-0.92)	-0.299
Credit	-0.006 (-0.06)	-0.004
Nonffarm	0.0282 (0.22)	0.018
Constant	-1.86 (-5.60)	
Dependent variable is Count		Number of obs = 178
Wald χ^2 (18) = 210.78		Prob χ^2 = 0.00
Loglikelihood = -162.72		Pseudo R^2 = 0.185

Notes: ***-Significant at 1%; **- Significant at 5%; and *- Significant at 10%. Values in parentheses are the ratio of the coefficient to the estimated asymptotic standard error. The method of estimation is Stata's Robust option following Huber/White standard errors and covariance.

The goodness of fit test results [(Goodness-of-fit χ^2 = 75.95 and Prob > χ^2 (159) = 1.00)] have given insignificant χ^2-tests indicating that the Poisson is appropriate model to explain variety count. The LR test has been used to check the significance of inclusion of a set of variables. The variables considered were classified into safety factors ($Risk_{proxy}$), concerns (nopurpose), endowment factors (lancrops, educate, plots, and totlabor), village dummies (Belinarba, Kerodeda, Asseliso, Ejeaneni, Chachole, and Gurbo) and access factors (access, cht'at, impexep, credit and nonffarm)[16].

[16]Variables are defined in Table 5.1.

The test fails to accept H_0 in all cases implying that all set of variables are important in their own right in explaining farmers' demand for variety diversification.

How are the results interpreted? Our point of interest is the expected count and interpretation should focus on the change in the expected number of FVs as either of the independent variables change. The simple way to interpret the results is by using the factor change. For instance, if age increases by 1 unit, the expected value of on-farm sorghum diversity increases by 0.72 % and if experience in growing IVs increases by 1 year, sorghum diversity on-farm decreases by 18.6 %. Coming to the dummy variables, being in Kerodeda increases on-farm sorghum diversity by a factor of 1.44 (= exp[0.363]) or increases by 44 percent (1.44 - 1), holding all other variables constant.

Most of the results confirm *a priori* expectations. They show that factors such as diversity of farmers' household needs being met by producing sorghum, farmers' sensitivity to income shocks, land heterogeneity, land size, and age are the factors significantly motivating farmers to manage a portfolio of local sorghum varieties. Through these in-built incentives, diversification on local varieties by farmers is expected to ensure the survival of at least some of the varieties under various environmental pressures.

Farmers with a variety of farm household concerns (*purposes*) have better propensity to more demand for local varieties. This implies that a single variety does not satisfy multiple purposes to which farmers wish to place the variety.

Farm households who fail to satisfy their subsistent needs from expected non-variate income and wealth have higher demand for multiple varieties of sorghum. They are trying to survive from the variety (ies) that tolerate the unexpected stress condition by planting as many varieties as they can.

In most applied work on crop diversity, fragmentation index (the ratio of the number of household plots to the total area of land operated) is used. However, it could be misleading because according to this index, a farm that has 2 plots with a size of a hectare each is equally fragmented (heterogeneous) with a single plot farm of half hectare. But heterogeneity is perhaps twice in the first than in the second. For this reason, number of plots and land size are used independently. Farmers who have reported large land holding have reported more number of local varieties *i.e.* an increase in land holding by one hectare increases the expected value of sorghum variety count by 16.7 percent.

Coming to the *dis-incentive* factors, access to extension and market, experience in using IVs and growing cash crops are the most important factors detaching the link between farmers' demand for sorghum diversity and their utility maximization. Farm households with better access to extension, markets and supply of IVs have reported less number of sorghum varieties on their farms. The utility that farmers derive from growing multiple varieties declines with the availability of these public goods. Similarly, as farmers get more experience in using IVs, the value of *poly-variety* to the household declines. These results warn us on the long-term potential of IVs and cash crop farming to replace FVs[17].

Ch'tat is a very important cash crop in the study area. As the involvement of the farm household in growing and marketing this crop increases, their demand for sorghum diversity declines. The financially non-rewarding sorghum is liable for replacement by cash crop farming as far as suitability of land does not hinder the replacement. The co-existence of crops of different rewards in a region, therefore, leads to endangering the diversity of the less important and less rewarding crops.

Expecting its positive impact on the use of IVs, education was predicted to have a negative effect on demand for sorghum count diversity. However, the variable has unexpected sign. Running a two-limit Tobit regression model on the proportion of sorghum plot allotted to IVs (Table IV-7 in Appendix IV-A) has shown that education attainment does not significantly affect use of IVs of sorghum. That could partly explain the unexpected sign in Table 5-2. Regarding the village dummy (*asseliso*), the sign contradicts with the simple descriptive statistics results probably because the sorghum diversity promoting factors are more prevalent in this village and there could be village differentiating variables not captured in the estimation.

The messages from the results of Table 5-2 above and Tables IV-3 and IV-4 (in Appendix IV-A) are pretty similar implying that variety count is a very good proxy to study farmers' derived demand for on-farm crop diversity.

---

[17]This result demonstrates one of the challenges in on-farm conservation programs and sends a pre-cautionary message of harmonizing crop diversity conservation and improved variety use aimed at meeting agricultural productivity objectives.

5.7. Summary and concluding remarks

This chapter has been about the key variables that characterize and induce farmers to contribute to *de facto* conservation of sorghum diversity. Most studies on crop diversity, especially those with ecological and botanical orientation, do not make a distinction among farmers. This chapter has given explicit account of farmers' heterogeneity (with respect to endowments, concerns and constraints) and its effect on poly-variety.

To alleviate problems of rural income polarization, poverty, and crop diversity loss, current strategies have to consider harmonizing agricultural productivity objectives with conservation objectives. In doing so, policy makers have to link incentives for maintaining FVs with incentives for using IVs. While the first calls for targeting pro-diversity farmers, the second requires targeting farmers with better comparative advantage to use IVs.

The results of this chapter by and large show that the poor are conserving sorghum diversity for free. If poverty reduction and food production strategies are targeting the poor in terms of improving their access to input supply (fertilizer and IVs), markets, agricultural extension and infrastructure development, such policies will have opportunity cost of reducing crop diversity on each farm over the long-term. Whenever hybrids are recommended for a certain locality, on-farm conservation strategies should also be part of the effort. The details of the conclusions and policy implications of the empirical results are presented in Chapter 9.

6. Farmers' coffee variety attribute preferences, demand for local varieties and incentives for poly-variety

6.1. General background

Undertaking on-farm conservation ventures requires understanding farmers' variety choice and variety attribute preferences. Such a study will also help in the areas of research priority setting and targeted adoption of IVs of crops. Thus, for a successful intervention (be it for adoption or on-farm conservation), policy has to be informed on:

- which farmers (commercial, subsistent, educated, poor, or risk takers) allocate the majority of their plot to local varieties;
- who prefers what kind of variety attributes most? and
- for which variety attributes are the improved varieties preferred? what about the local varieties?

The supply of variety attributes from which farmers choose is constrained by the genetic diversity of the crop in their community (Smale *et.al*, 2001). Even though the choice set (coffee varieties) that farmers have are the same within a PA, some farmers entirely plant local coffee trees, others both and still others only IVs. Coffee diversity produced on-farm also varies across farmers. As presented in Chapter 3, some farmers give most priority to yield, others to yield stability and still others to marketability. What explains such differences across farmers? What can policy learn from the results of this exercise?

Understanding farmers' variety attribute preferences will also be useful to predict the likelihood of survival of varieties embedding those attributes. This enables policy makers to identify varieties for which policy incentives are required and those that can be maintained *de facto*. In agricultural research priority setting too, understanding farmers' variety attribute preferences will serve as an input for developing varietal technologies with more chance to be adopted and be successful (Adesina and Baidu-Forson, 1995; Adesina and Zinnah 1993).

This chapter examines farmers' effective demand for coffee variety attributes, potential demand for multiple local coffee varieties, effective demand for multiple variety attributes, and effective demand for coffee diversity using a variety of indicators for the respective variables. The specific objectives to be addressed in this chapter include:

- studying farmers' preferences for coffee variety attributes and their diversification behavior on coffee variety attributes;
- identifying the household contextual factors affecting farmers' effective demand for local coffee trees on-farm and their incentives to undertake *poly-variety*; and
- identifying the varieties that will be conserved *de facto* and those that need policy incentives.

The remaining part of this chapter is organized as follows. The next section presents the linkages among farmers' concerns, variety attribute preferences and land allocation. The theoretical framework is contained in Section 6.3. The econometric models employed are subsequently presented in Section 6.4. The variables employed in the econometric analysis are then described in Section 6.5. Section 6.6 presents regression results and related discussions. The last section summarizes the chapter.

6.2. Farmers' concerns, variety attribute preferences and land allocation

Farmers have economic (resource constraints, markets and risk) and non-economic (religion, culture and norms) concerns. Of course, not all farmers' motives are purely economic or non-economic, although some are more relevant than others for economic behavior (Gasson, 1973). When local variety attributes satisfy farmers' concerns, their *de facto* conservation is the outcome of the harmony of variety attributes and farmers' concerns. Thus, the level of *de facto* conservation that occurs in an area is mainly a function of the capacity of FVs to satisfy farmers' economic and non-economic concerns. In essence, farm household characteristics and variety characteristics translate to variety attribute preferences and land allocation decisions.

Most of the risk cautious farmers are looking for combinations of varieties that enable them to satisfy subsistent needs. They are not interested in a variety that is best either for yield or for resistance to disease or early maturity. For such farmers, a variety, that is an outlier in a single variety attribute, may not be of interest.

Farmers have multiple concerns and no single variety satisfies all their concerns (Brush, 2000; Bellon, 1996). As the number of concerns and desires of farmers increases, the criteria they use to make on-farm variety use decision increases (Teshome *et.al*, 1999). Having this constraint, most farmers are managing diverse set of local varieties of crops to get as many desirable attributes as possible. This is what is called *poly-variety* (Netting and Stone, 1996). A critical issue that remains to be studied is how farmers rank their concerns and how they may trade-off one against another (Bellon, 1996). Their variety attribute preferences are the reflections of their concerns and hence studying variety attribute preferences implies studying their concerns indirectly.

Farmers can be grouped by the concern profiles they share. That, in turn, should be related to the environment in which they farm and the degree of market integration (Bellon, 1996). A study by Bellon (1991) has grouped farmers' concerns into 3: *agro-ecological* (rainfall and soil type); *technological* (response to fertilizer, labor and cultural practices); and *use* (performance of the variety with respect to the purposes and uses of the output such as food taste, market price, yield, and production of straw or fodder).

Why are we concerned with farmers' variety attribute preferences and concerns? Understanding farmers' variety attribute preference serves as an input to both on-farm conservation and technology development and diffusion. For on-farm conservation, it serves in the areas of identifying farmers to be targeted and varieties maintained *de facto* and those requiring external policy incentives. Contextual variety development, breeding priority setting, targeted diffusion of IVs, and better potential of successful outcome are the benefits from the technology side.

Farmers with similar concerns should have similar variety attribute preferences. Their incentives and/or dis-incentives to use certain varieties of crops will be the result of the harmony of household concerns and variety attributes. This, in turn, implies that for farmers with similar household concerns, similar set of on-farm conservation policy incentives should convince them. Figure 6-1 sketches the relationships among farmers' concerns, their contextual characteristics, environmental characteristics, and crop diversity. As presented in the scheme, farmers' concerns are the outcomes of their characteristics, the policy environment, and environmental characteristics in which they are operating. Addressing concerns requires matching variety attributes with farmers' concerns. This impings on farmers' variety management that includes selection among the available choice set and use decisions (mixing, land allocation, input use, cultural practices and so on). The farmers' variety choice and management decisions will then determine the outcome variable: crop diversity.

Figure 6-1: A simplified conceptual characteristic model

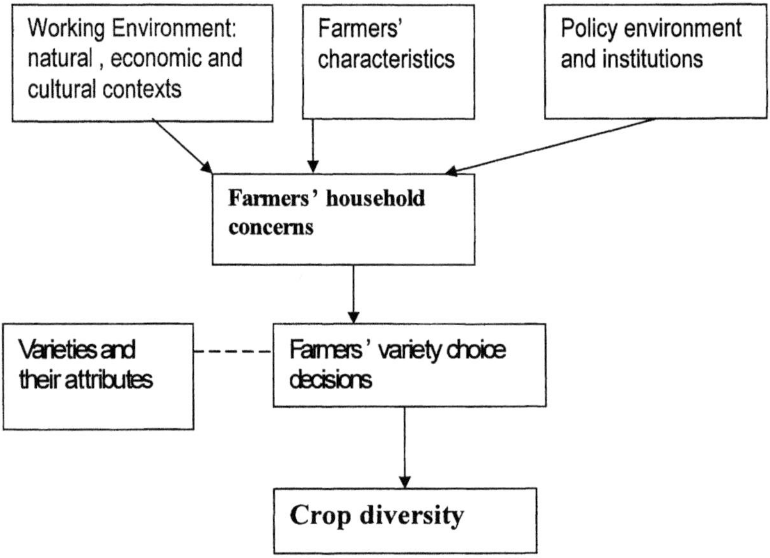

Each farmer derives different utility from consuming different variety attributes based on the concerns facing him/her. Different farmers will rank variety attributes differently based on their concerns, resource endowments, and constraints. Thus, a variety that is not of interest for farmer 'A' may be maintained by farmer 'B' and *vice versa*.

The probability of existence of a variety on farmers' fields is a function of the extent to which it embeds the important attributes playing a key role to the household. Thus the question boils down to the fitness of the variety attributes with household concerns. The survival of a variety on-farm is dependent on its capacity to supply the variety attributes that receive more weight by the farm household.

Due to the complexity of farmers' variety use decisions, the micro-economic model should be utility-based as a profit maximization framework is unable to explain farmers' variety attribute preferences and their land allocation decisions. The following section attempts to derive attribute-based diversity and variety attribute preferences from farmers' utility maximization.

6.3. The theoretical framework

The theory presented here draws from Lancaster's (1966) characteristic model, Roy's (1952) safety first model, and the random utility theory (McFadden, 1974). The partial adoption model (Feder *et.al*, 1985 and Feder and Umali, 1993) and agricultural household models (Singh *et.al*, 1986) are the analytical foundations for relating the conceptual framework with the estimated econometric models.

Variety portfolio enables farmers to consume different set of attributes from different varieties. Through their derived demand for varieties, farmers are actually consuming variety attributes and producing crop diversity while producing crops to satisfy household needs. However, they are neither directly paying (except the opportunity cost involved) nor being paid for their contribution in maintaining crop diversity on their farms.

In a neoclassical model of profit maximization, an unconstrained risk neutral farmer who maximizes profits would choose to grow only the variety with the highest profits per hectare (Smale *et.al*, 2001). This is true under very restrictive technology assumptions: *non-jointness* of inputs and outputs and *constant returns to scale*. Coffee farmers can not specialize in a single variety attribute (like yield) because they have other non-yield household concerns and risk behavior. Above all, there are no *constant returns to scale* for every variety.

At any point in time, the farm household has a chance to replace certain coffee trees with new seedlings. The old coffee trees and the new ones are judged by their respective capacity to satisfy consumption and production attributes demanded in line with farm household objectives, values, endowments, and constraints (x_i's).

Suppose Q_p, Q_p-Q_s, Y_f and $C_1 \ldots C_n$ denote quantity of coffee produced, quantity of coffee consumed, farm income and household consumption of other on-farm produced or purchased goods, respectively. Assuming a well behaved utility function, the farmers' utility maximization problem can be set as:

$$Max: \; U = U(Q_p - Q_s, C_1, \ldots, C_n / \varpi_1) \qquad \text{Household utility maximisation (6.1)}$$

where ϖ_1 refers to the vector of household level contextual factors affecting farmers' utility maximization[1].

The utility is subject to cash, labor, land, and production technology constraints as presented in Chapter 5. Following the conventional utility maximization framework, one can derive the following constraint:

$$p_c(Q_p - Q_s) - c(Q_p;\varpi_1) + Y_{NF} \le Y \qquad \text{Full income constraint (6.2)}$$

where p_c, Q_p, $c(.)$, Y_{NF} and Y refer to price of coffee, total coffee produced (aggregating all coffee types for simplification), cost of production for coffee, income from other enterprises and transfers, and Becker's concept of full income, respectively. The production technology influences the relative importance of the different variety attributes and thereby the varieties of coffee to be maintained depending on the attributes embedded in each. Assuming perfect markets and risk free environment, the first order conditions of this problem (maximizing utility subject to production and full income constraints) lead, among other things, to the optimum level of employment of the respective inputs and coffee production (C^*_p).

From the level of production inputs used, farmers' variety attribute preferences, effective demand for multiple variety attributes, effective demand for local varieties, and effective demand for poly-variety can subsequently be derived by tracing the attributes of varieties preferred, variety attribute count in farmers' choice basket, the proportion of land allocated to FVs, and the number of FVs under cultivation. These are all derived outcomes of farmers' *revealed preferences*.

For the sake of convenience, consider as if the household derives utility only from coffee. With this assumption, the different varieties of coffee are the different enterprises. Thus, the derived outcomes are '*n*' varieties of coffee ('*n*' enterprises), greater than or at least equal to '*n*' variety attributes embedded in the variety (ies) planted, and '*n*' proportions of land allocated to each enterprise. Having 'n' varieties of coffee in a given locality embedding 'n' or more variety attributes, the point of interest is the number of varieties or variety attributes in each farmer's choice basket, the proportion of local varieties on each farm, and/or the ranking and preferences of variety attributes.

---

[1]See Section 5.6.for details of the variables considered.

Given that different varieties have different input requirement and marketability, the input and output market constraints will force farmers to accordingly adjust their *revealed preferences* for varieties and their attributes. This effect, in turn, will change the relative importance of different coffee variety attributes to the household and farmers' land allocation decisions. Thus, the following hypotheses can be formulated in this connection:-

- While yield stability and environmental adaptability are important attributes for farmers with little access to markets, yield and marketability are the most important variety attributes for farmers found in more accessible areas; and
- Farmers less integrated to markets have higher demand for multiple coffee variety attributes and higher effective demand for local varieties of coffee.

Farmers' preferences are driven by the need to optimize on the use of scarce factors of production to save these inputs (Pingali *et.al*, 2001). This is consistent with the *induced innovation hypothesis* (Hayami and Ruttan, 1985). Accordingly, trends and patterns of input and output market prices will impact on the type, number and relative share of FVs in use. Thus, the other hypotheses to be tested will be:

- Land and labor constrained farm households have lower demand for multiple coffee variety attributes;
- Farmers' effective demand for local varieties increases with capital constraint; and
- Land and labor constrained farm households have higher demand for yield and marketability.

The next most important variable worth considering is risk. To study the linkage between farmers' variety diversification and risk, the notion of *Roy's safety first* model is integrated into the utility-based derivation of farmers' preference for variety attributes and coffee land allocation decisions.

The essence of the safety-first strategy is assuming a *lexicographic optimizer i.e.* an economic agent who aims at meeting the target minimum survival level as first priority objective and maximizes expected returns as second priority. In economic theory, lexicographic preferences stand for a situation whereby one good (survival in our case) has such a high priority in an individual's preference ordering and needs to be satisfied first before an increase in farm income can lead to an increased utility. In a sense, the *safety first principle* follows the *precautionary principle* which involves taking action before facing any potential risk.

Concerning coffee farming which does not give harvest the first planting year, those farmers who are far below the subsistence requirement will be less inclined to replace existing local coffee trees. Otherwise, they will have to wait 3-4 years to harvest from the new coffee trees which they can not afford due to their immediate consumption requirement. For the purpose at hand, FN_{income} (non-covariate income and wealth) for each farm household is computed as the sum of annual estimated income from non-farm, off-farm and unearned income sources (Y_{NF}) and the value of livestock (V_{live}) *i.e.*

$$V_{NF} + V_{live} = FN_{income} \qquad \text{Expected income and wealth (6.3)}$$

FN_{income} and the minimum household requirement are used to compute the risk proxy as the difference between FN_{income} and the minimum requirement (see Section 5.3 for details). Based on the suitability of variety attributes to address farmers' risk concern, it can be hypothesized that:

- yield stability and environmental adaptability are important variety attributes for income-shock vulnerable farmers and yield and marketability are the most important variety attributes for income-shock tolerating ones; and
- income-shock vulnerable farmers have higher demand for multiple coffee variety attributes and higher effective demand for local varieties of coffee[2].

Not only is farmers' preference for variety attributes influenced by their input constraints and concerns but also are their preferences for variety attributes important for their land allocation decisions among different varieties. Thus, depending on the features of local varieties, it can be hypothesized that:

---

[2]Those farmers who are more vulnerable to potential income shocks are hypothesized to diversify on their demand for coffee variety attributes because they are demanding more production attributes unlike the less vulnerable farmers who will specialize in one or two most important variety attributes. Income-shock vulnerable farm households do not look for a variety that is exceptionally superior in a single attribute like yield but a variety which has many desirable attributes.

- farmers with higher preference for yield and marketability have lower demand for local varieties of coffee while those with higher preference for yield stability and environmental adaptability have higher effective demand for local varieties; and
- farmers' incentive to undertake *poly-variety* increases with their demand for yield stability and environmental adaptability and decreases with their demand for yield and marketability.

In what follows, the specific econometric models to be estimated are described.

6.4. The econometric models

6.4.1. The MNL model

Consider two bundles of varieties *i.e.* FVs and IVs. The FVs contain Z^{1}_{FV} ....Z^{n}_{FV} attributes and the IVs contain $Z^{1}_{IV} \ldots Z^{n}_{IV}$ attributes. Farmers allocate their land among different varieties according to the shadow price of the attributes embedded in the respective varieties. Taking the different varieties as intermediate inputs and the attributes contained therein as final products it can be underscored that farmers are ultimately consuming variety attributes. Hence, variety attributes will be put in the direct utility function as ingredients.

Suppose Z_i^j is a vector of attributes of coffee varieties in the choice set and ϖ_1 is the vector of characteristics of farmers reflecting their endowments, concerns and preferences. Then utility from coffee can be given as a function of the optimum set of attributes contained in the coffee varieties under production *i.e.*

$$U_{coffee\ production} = f(Z^{1}_{FV} \ldots Z^{n}_{FV}, Z^{1}_{IV} \ldots Z^{n}_{IV} / \varpi_1) \qquad \text{Optimum No.of coffee attributes} \quad (6.4)$$

From the available choice set, the ith farmer is assumed to select the combination of variety attributes that will maximize the utility derived. If there are 7 desirable and 3 undesirable variety attributes contained in the total available coffee varieties in a certain village, the random utility theory predicts that rational farmers rank all the possible combinations and take the ones generating maximum utility subject to their constraints.

According to Lancaster's *characteristic model* (Sub-section 4.6.3), utility is derived from the attributes contained in the selected varieties. The variety attributes in the farmers' choice basket are the sub-set of the total choice set.

Suppose $U_{ij} = \mu_{ij} + \varepsilon_{ij} = X'_{ij}\beta + \varepsilon_{ij}$, where μ_{ij} is a non-stochastic function of observables and ε_{ij} is an unobservable stochastic error term representing the combined effects of unobserved factors and random individual behavior. If option 1 (j = 1) with the corresponding utility $U_{i1} = \mu_{i1} + \varepsilon_{i1} = X'_{i1}\beta + \varepsilon_{i1}$ is chosen, then according to the random utility model, U_{i1} is the best utility that can be attained.

Suppose there is a chance for farmers to choose among M variety attributes as a choice set, indexed j=1,2,…,M, noting that the order is arbitrary. There will, then, be a corresponding utility level (U_{ij}, j=1,2,…,M) attached to each variety attribute or combination of variety attributes. Let the probability that the ith farmer chooses the jth variety attribute be P_{ij} and denote the choice of the ith farmer by $Y'_i = (Y_{i1}, Y_{i2}, \dots, Y_{ij})$ where Y_{ij} = 1 if the jth attribute is selected and all other elements of Y'_i are zero. Provided that each farmer is observed only a single time and he/she has a single choice among the available choice set, the likelihood function of the sample of values Y_{i1}, …, Y_{ij} is:

$$L = \prod_{i=1}^{n} P_{i1}^{Y_{i1}} P_{i2}^{Y_{i2}} \dots P_{ij}^{Y_{ij}}$$ The likelihood function for MNL (6.5)

Using the maximum likelihood optimisation procedure and assuming that each farmer is maximizing utility, the probability that farmer i chooses variety attribute 1 is given by:

$$P_{i1} = \Pr\lfloor U_{i1} > U_{i2},\ U_{i1} > U_{i3}, \dots, \text{ and } U_{i1} > U_{ij} \rfloor$$ Attribute choice (6.6)

The probabilities of the remaining variety attributes can be expressed using similar relationships. The probability that the household chooses the variety attribute (j) that maximizes the household utility can be given as:

$$P\{y_i = j\} = P\{U_{ij} = \max\{U_{i1}, \dots, U_{ij}\}\}$$ Random utility (6.7)

Taking exp(μ_{it}) in stead of μ_{it} to ensure non-negative probability leads to:

$$\Rightarrow P\{\mu_{it} + \varepsilon_{it} > \max_{k=1,\ldots,j,k\neq l}\{\mu_{ik} + \varepsilon_{ik}\}\}$$

$$\Rightarrow \frac{\exp\{\mu_{it}\}}{\exp\{\mu_{i1}\} + \exp\{\mu_{i2}\} + \ldots + \exp\{\mu_{ij}\}}$$ Random utility (6.8)

Writing $\mu_{it} = x_{it}^{'}$, constraining one of the vectors of β's to zero (parameter vectors for disease resistance in our case) for the sake of identification[3] and assuming that the errors (ε_{ij}) are independently and identically distributed leads us to the following multinomial logit (MNL) model:-

$$\Pr\{y_i = j\} = \frac{\exp\{x_{it}^{'}\beta\}}{1 + \exp\{x_{i2}^{'}\beta_2\} + \ldots + \exp\{x_{ij}^{'}\beta_J\}} = \frac{\exp\{x_{ij}^{'}\beta\}}{1 + \sum_{j=0}^{3}\exp\{x_{ij}^{'}\beta_J\}}$$ MNL model (6.9)

where j = 0 for yield (Y); j = 1 for yield stability (YS); j = 2 for environmental adaptability (EA); j = 3 for marketability (M); and j = 4 for disease resistance (DR). Assuming that disease resistance is taken as a reference group, this model equates the probability that the farmer gives prior preference to variety attribute j *i.e.*

$$P\{y_i = Y\} = \frac{\exp\{x_{ij}^{'}\beta\}Y}{1 + \exp\{x_{ij}^{'}\beta\}Y + \exp\{x_{ij}^{'}\beta\}YS + \exp\{x_{ij}^{'}\beta\}EA + \exp\{x_{ij}^{'}\beta\}M}$$ Prob. that the

farmer opts for yield (6.10a)

$$P\{y_i = YS\} = \frac{\exp\{x_{ij}^{'}\beta\}YS}{1 + \exp\{x_{ij}^{'}\beta\}Y + \exp\{x_{ij}^{'}\beta\}YS + \exp\{x_{ij}^{'}\beta\}EA + \exp\{x_{ij}^{'}\beta\}M}$$ Prob. that the

farmer opts for yield stability (6.10b)

$$P\{y_i = EA\} = \frac{\exp\{x_{ij}^{'}\beta\}EA}{1 + \exp\{x_{ij}^{'}\beta\}Y + \exp\{x_{ij}^{'}\beta\}YS + \exp\{x_{ij}^{'}\beta\}EA + \exp\{x_{ij}^{'}\beta\}M}$$ Prob. that the

farmer opts for environmental adaptability (6.10c)

[3]The choice is arbitrary.

$$P\{y_i = M\} = \frac{\exp\{x_{ij}^{'}\beta\}M}{1+\exp\{x_{ij}^{'}\beta\}Y+\exp\{x_{ij}^{'}\beta\}YS+\exp\{x_{ij}^{'}\beta\}EA+\exp\{x_{ij}^{'}\beta\}M}$$ Prob. that the farmer opts for marketability (6.10d)

As the probability of all the options can not exceed unity, subtracting the sum of 6.10a to 6.10d from 1 results in:

$$P\{y_i = DR\} = \frac{1}{1+\exp\{x_{ij}^{'}\beta\}Y+\exp\{x_{ij}^{'}\beta\}YS+\exp\{x_{ij}^{'}\beta\}EA+\exp\{x_{ij}^{'}\}M}$$ Prob. that the farmer opts for disease resistance (6.10e)

Its log-likelihood function can, thus, be given as:

$$l\{\beta_1,...,\beta_j\} = \sum_{j=0}^{4}\sum_{i=1}^{n} x_{ij}\beta - \sum_{i=1}^{n}\log\left(1+\sum_{j=0}^{4}\exp(x_{ij}\beta)\right)$$ Log-likelihood for MNL (6.11)

The MNL model is used to predict the probability that a farmer prefers a certain variety attribute and how that preference is influenced by different household and environmental related factors. This modeling exercise has enabled us to answer the question '*Who prefers what*?' '*What kind of farmers are looking for what types of varieties?*'

The respective signs of the coefficients in the MNL model do not necessarily imply the signs of the marginal effects. The sign of the marginal effect depends not only on the respective coefficients but also on the relative size of the expected value of the coefficients across the choices and the value of the coefficient in the choice set (Greene, 2000) *i.e.*

$$\frac{\partial P_j}{\partial X_j} = P_j\left[\beta_j - \sum_{k=0}^{j} P_k\beta_k\right] = P\left[\beta_j - \overline{\beta}\right]$$ Marginal effects for MNL (6.12)

Thus, in the case of MNL model, one can not for sure tell the sign of the relationship based on the coefficients until the marginal effects are computed. The parameters provide the change in the log of odds ratio (the ratio of P_i to 1 - P_i) per unit change in the explanatory variable, not the change in the probability itself. For this reason, both the coefficients and the marginal effects are presented in Section 6.6.

MNL assumes that the unobserved factors are uncorrelated over alternatives and they have the same variance for all alternatives (Train, 2003). This is the most important limitation of this model that has to be taken into account in interpreting the regression results.

6.4.2. The Logit transformed OLS regression

Number of local varieties on a farm does not tell us farmers' demand for local varieties as it may not signify the relative importance of the respective set of varieties (improved and local) on the farm. It may be that 2 IVs cover much more area than 4 FVs on a farm. For this reason, the proportion of local coffee trees on farmers' fields allocated to coffee is used as a proxy for farmers' demand for local varieties.

Suppose $P_{FV} = (P_1, P_2, P_3, \ldots, P_n)$ is a vector of proportions of land use reflecting the allocation of land to local varieties on-farm. As the member of P_{FV} on each farm increases, the utility derived from FVs also increases. This increases the probability of survival of FVs on the respective farm.

Considering P_{FV} as a land allocation decision, the proportion in reduced form (P_{FV}) can be given as a function of factors affecting farmers' land use decisions (ϖ_1):

$$P_{FV} = f(\varpi_1) \quad \text{Explaining the proportion data (6.13)}$$

P_{FV} can also be taken as a crop diversity indicator based on the premises that the larger the proportion of indigenous trees on-farm, the more will be the likelihood of higher coffee diversity on-farm. Cromwell and van Oosterhout (2000), among others, have made this assumption while using the proportion of the total farm area allocated to growing small grains as diversity indicator.

Following logit transformation procedure, transformation of the dependent variable is performed first *i.e.*

$$TransP_{FV} = \ln\left(\frac{P_{FV}}{1 - P_{FV}}\right) \quad \text{Logit transformation (6.14)}$$

Then the transformed variable ($TransP_{FV}$) is used as a response variable to conduct OLS regression. Breusch-Pagan test for heteroscedasticity has failed to reject the OLS model with homoscedasticity (χ^2_{21} *tabulated* = 32.7 and χ^2 *calculated* = 31.14). For this reason, the OLS results reported in Table 6.8

below are without correcting for heteroscedasticity. The model estimated is set as:

$$TransP_{FV} = \beta_0 + \beta_1 x_{i1} + \beta_2 x_{i2} + ... + \beta_k x_{ik} + \varepsilon_i \text{ Logit-transformed OLS} \quad (6.15)$$

where i takes values from 1 to n, k is the number of regressors, ε is the stochastic error term, β's are the parameters estimated, and x_i's are the variables described in Section 6.6 (Table 6-3).

The other important option as a cross-check has been two limit (zero and one) tobit model (following Smale *et.al*, 2001) due to the nature of the distribution of the data (1, 0 extremes and continuous values in between). Comparing Tobit with heteroscedasticity and without heteroscedasticity using LR-test has failed to reject the Tobit regression model with homoscedasticity (χ^2_{21} *tabulated* = 41.4 and χ^2 *calculated* = 4) and hence the results reported in Appendix IV-B (Table IV-9) are without correcting for heteroscedasticity.

6.4.3. The Poisson regression

Farmers' demand for diversity is examined counting the number of attributes that farmers have mentioned to be important to their households. This diversity index, called *attribute count*, is newly proposed. The number of FVs that farmers are maintaining is the second response variable to analyze farmers' effective demand for *poly-variety* as it is done in Chapter 5. Considering crop diversity as a positive externality of farmers' utility maximization[4], coffee diversity is given as:

$$D_{c1} = (\sum_{i=1}^{n} v_i / \varpi_1) \quad and \quad D_{c2} = (\sum_{i=1}^{m} T_i / \varpi_1) \text{ Variety count and attribute count} \quad (6.16)$$

where D_{c1} is variety count index where v_i's (v_i = 1 if the variety is selected and zero otherwise) stand for FVs and D_{c2} is attribute count index where T_i's (T_i = 1 if the variety attribute is important to the household and zero otherwise) stand for coffee variety attributes found to be relevant to the household. '*n*' and '*m*' are the maximum number of FVs and variety attributes the village is endowed with, respectively. They stand for farmers' choice set in a village. The variables D_{c1} and D_{c2} are explained using the

[4]See Chapter 5 for details.

Poisson regression model[5]. When regression-based tests of over-dispersion were run, the variables '*locattribute*' and '*localvar*' (See Table 6-2 for description) were found to be under-dispersed. Hence, Stata's robust option was used to deal with the violation of the equi-dispersion assumption of the Poisson regression model like in Chapter 5.

The proxy D_{c2} is taken as diversity indicator based on the premise that the larger the number of desirable coffee variety attributes important to the farm household, the larger will be the amount of coffee diversity maintained on-farm as a single variety is unlikely to address all farmers' preferred attributes (Bellon, 1996). On the contrary, if farmers are exclusively looking for a single variety attribute, like yield, then, *ceteris paribus*, they will not have incentive to maintain more than one variety.

Due to farmers' inability to exhaustively list all varieties, D_{c1} has not been a good proxy as the regression results will reveal. As a result, the index D_{c2} is preferred. In using D_{c1} and D_{c2} as proxies for coffee diversity, a diversity of zero is assigned for all IVs.

6.5. Description of variables

Based on the theoretical framework presented in Section 6.3, the factors (ϖ_l) considered to study the response variables of our interest are household characteristics (experience in farming, education level of the household head, farmers' experience in growing IVs, relative importance of coffee to the household[6], and village dummies), resource endowment (presence of income outside agriculture, labor endowment, land endowment, and land heterogeneity), household concerns (capital shortage, lack of government services – credit, extension, and input supply; natural factors – pests, disease, weather, and drought), farmers' sensitivity to potential income shocks, access to markets, and farmers' comparative assessment of FVs and IVs with respect to different variety attributes.

The average number of local varieties of coffee reported is 1.28 with a minimum of 1 and maximum of 3. Among the coffee growing farmers

---

[5]The suitability of this model for cout data and its major ingredients are contained in Chapter 5.

[6]If a farmer is planting and getting income only from coffee, then, *ceteris paribus*, the number of attributes of coffee that he/she will find important to decide on use and replacement of coffee and the intensity of his sensitivity to any income shocks from coffee is expected to be high.

interviewed, about 74 percent of them have reported to have a single local variety. However, farmers' inability to name all local varieties has to be noted. Most farmers could not identify names of local coffee varieties. Mostly, they put the varieties into two categories – '*begeja*' (local name for the indigenous coffee trees) and *'project'* (local name for the improved coffee trees).

Tables 6-1 to 6-3 below define the variables used in the regression, provide the expected signs based on the theoretical predictions along with the descriptive statistics for the respective variables.

Table 6-1.Variable definitions and expected signs: farmers' preference for yield and marketability

Variable	Description	Mean	SD	Expected sign
Attribute	0 –Yield; 3 – marketability			DV
Yildrank	Farmers' preference ranking for yield (the lower the better)	2.17	0.89	DV
Mrktrank	Farmers' preference ranking for marketability (the lower the better)	3.14	1.14	DV
Explanatory variables				
Howlong	The number of years the farmer has been planting the local coffee variety (ies)	24.32	13.96	+
Age	Age of the household head	46.3	13.7	-
Agesqur	The square of 'age'	2332.5	1383.5	-
Educathh	Education level of the household head	2.98	3.20	+
Anyincom	Does the household have any income source outside agriculture? (1 – yes; 0 – no)	0.25	0.43	-/+
Accesse2	The average of walking time needed to reach the extension agent, the nearest weather road and market (average sum to the three in minutes)	30.49	22.03	-
Timarket	The average of walking time needed to reach the nearest market (in minutes)	44.67	33.03	-
Cofland	Proportion of land allocated for coffee as a percentage of total land holding	0.54	0.30	-/+
Labindic	Labor shortage indicator – calculated as a ratio of full	1.18	1.19	-

	time labor equivalent to land size			
Labor14	Farmers' ranking of the importance labor shortage as a production constraint (the lower the better)	5.4	2.9	-
Lanindic	Land shortage indicator – calculated as a ratio of land size to household size in consumption need equivalent	2.03	3.60	-
Land14	Farmers' ranking of the importance of land shortage as a production constraint (the lower the better)	3.3	2.7	-
$Risk_{proxy}$	Risk proxy as defined in equation 5.16	-546.24	659.46	+
Capital4	Farmers' ranking of the importance of working capital as a production constraint (the lower the better)	2.41	1.75	-
Govrank	Farmers' ranking of the importance of extension, input supply, and farm implements as production constraints (the lower the better)	5.78	2.14	-
Naturank	Farmers' ranking of the importance of natural factors (pests, disease, weather, and drought) as a production constraint (the lower the better)	3.91	1.64	-/+
Harro[7]	Village dummy (for Harro)	0.14	0.35	-/+
Kelaguda	Village dummy (for Kela Guda)	0.13	0.33	-/+
Kelokiri	Village dummy (for Kilole Kirkir)	0.13	0.33	-/+
Gibeboso	Village dummy (for Gibe Boso)	0.11	0.31	-/+
Halosebe	Village dummy (for Halo Sebeka)	0.15	0.35	-/+
Sebekdeb	Village dummy (for Sebeka Debiye)	0.14	0.34	-/+

Notes: DV – Dependent variable

[7]Villages – Bulbulo and Yachi Urechi – are left as a reference.

As given in the table, the dependent variable '*attribute*' takes five discrete values (0 – yield, 1 – yield stability, 2 – environmental adaptability, 3 - marketability, and 4 – disease resistance). Disease resistance is taken as a reference in the regression. In contrast to attributes presented, the values of '*attribute*' designated by 1 and 2 are hypothesized to have the opposite of the sign presented in the table unless it happens to be unpredictable.

To elucidate farmers' demand for aggregate variety attributes, variety attributes important to coffee farmers in their decisions to maintain local coffee trees were first identified from the key informant preliminary interviews[8]. Following this, coffee-growing farm household heads were provided with the following query: "Which of the following variety attributes are important for your decision to maintain local variety (ies) of coffee on the farm?"

a. Yield ___ b. Yield stability___ c. Very good harvest during a good season ____ d. At least some harvest during a bad season ___ e. Fertilizer responsiveness ____ f. Tolerance to environmental stress – drought, frost, wind, soil fertility ____ g. Disease resistance ____ h. Marketability ____i. Taste ____j. Early maturity ____ k. Other attributes (please specify) ____"

For all the survey households, an average of 3.7 desired attributes for coffee varieties were mentioned. The results show that 37, 48, 56, 37, 23, 37, and 28 of the farmers have reported 1, 2, 3, 4, 5, 6, and 7 attributes, respectively.

Table 6-2. Variable definitions and expected signs: farmers' potential and effective demand for coffee diversity

Variable	Description	Mean	SD	Expected sign
Localvar	The number of local varieties of coffee grown on the plots	1.28	0.50	DV
Locattribute	Number of variety attributes mentioned to be important to the household (for local variety use)	3.69	1.92	DV
Explanatory variables				
Educathh	See Table 6-1			-
Anyincom				-
Accesse2				+
Timarket				+

[8]This was done also to solicit the most important variety attributes farmers are concerned with and the major production constraints in the respective villages.

Experenc	Experience in farming	22.3	12.0	+
Expsqur	The square of '*experenc*'	642.1	645.3	+
Labshort	HHs whose major concern is labor shortage (taking those who gave rank 1 or 2)	0.17	0.37	-/+
Labendow	Number of full time workers in the household	4.25	1.99	+
Capshort	HHs whose major concern is capital shortage (taking those who gave rank 1 or 2)	0.68	0.47	+
Capital4	See Table 6-1			+
Govshort	Households whose major concern is lack of government services (extension, markets, and input supply) (taking those who gave rank 1 or 2)	0.30	0.46	+
$Risk_{proxy}$	See Table 6-1			-
Govrank				-
Naturank				-/+
Natshort	HHs whose major concern is natural problems (taking those who gave rank 1 or 2)	-/+	0. 43	-/+
Landsize	Land size operated by the household (Fechasas[9])	6.39	5. 57	+
Purposes	Number of purposes of growing coffee	2.12	1.04	+
Plots	Number of plots per household	2.43	1. 34	+
Yildrank	Farmers' ranking for yield (the lower the better)	1.17	0.89	+
Stabrank	Farmers' ranking for yield stability (the lower the better)	1.95	1.27	-
Adaprank	Farmers' ranking for environmental adaptability (the lower the better)	2.17	1.13	-
Mrktrank	Farmers' ranking for marketability (the lower the better)	2.14	1.14	+
Diserank	Farmers' ranking for disease resistance (the lower the better)	2.66	1.33	-/+
Village dummies	See Table 6-1			-/+

Notes: DV – Dependent variable

[9]Fechasa is a local unit of measurement which equals about one fourth of a hectare.

Table 6-3. Variable definitions and expected signs: farmers' effective demand for local varieties of coffee

Variable	Description	Mean	SD	Expected sign
Logtree	Log-transformed '*Proptres*'	3.0	4.6	DV
Proptres	Proportion of local coffee trees in the field (as a percentage of total trees)	0.69	0.30	DV
Explanatory variables				
Experenc	See table 6-1			+
Educathh				-
Accesse2				+
Lanindic				-/+
Labindic				-/+
Regular				-
$Risk_{proxy}$				-
Capital4				-
Govrank				-/+
Naturank				-
Beteryil	Ranking local and improved varieties in terms of yield (0 – IVs are better; 1 – both are equal; 2 – FVs are better)	0.55	0.64	-
Betedrot	Ranking local and improved varieties in terms of drought tolerance (0 – IVs are better; 1 – both are equal; 2 – FVs are better)	1.21	0.75	+
Betestab	Ranking of improved and local varieties in terms of yield stability (0 – IVs are better; 1 – both are equal; 2 – FVs are better)	0.65	0.66	+
Betemakt	Ranking of improved and local varieties in terms of marketability (0 – IVs are better; 1 – both are equal; 2 – FVs are better)	0.99	0.40	-
Beterfod	Ranking of improved and local varieties in terms of food taste (0 – IVs are better; 1 – both are equal; 2 – FVs are better)	1.20	0.56	+
Village dummies	See table 6-1			-/+

Notes: DV – Dependent variable

The average size of the land reported by farmers to have been operated is 6.39 Fechasas. Only 24.53 % of the sampled households have reported to have earned income outside agriculture. The average grade of schooling for the heads of the sampled households is about 3 years. While the average number of years coffee farmers keep local varieties on-farm is about 24 years, their farming experience, on average, is 22 years. Thus, most farmers' experience with farming is equivalent to their experiences with growing coffee.

The walking distance of the farm household's place of residence from a market (in minutes) is used as an indicator for market access. Farmers need, on average, about 30 minutes walk to reach the nearest market, dry weather road and extension agent.

The mean $Risk_{proxy}$ is negative implying that, on average, all households can not satisfy $Basic_{req}$ if non-farm income and value of livestock are equally re-distributed. This is in contrast to the result presented in Section 5.5 for sorghum growing farmers. The major reason could be cht'at and there could be other income and wealth items for coffee growing farmers not accounted for.

The relative importance of coffee to the sampled farmers is so high that they are, on average, allocating about 54 percent of their land to coffee. Land shortage is a first priority concern for 44 percent of the households. Labor shortage is a priority concern only for 17 percent of the coffee farmers while capital is for 68 percent of the farm households. Surprisingly enough, only 30 percent of the farm households have ranked lack of services (extension, markets and prices) as a priority concern.

To elucidate farmers' derived demand for variety attributes, they were asked to choose among alternative variety attributes (See Question 23.1 of the Questionnaire in Appendix V). The descriptive statistics show that 29.3, 29.3, 21.8, 13.9, and 5.64 percent of the farmers had opted for yield, yield stability, environmental adaptability, marketability, and disease resistance, respectively. More details about the data generating process and data description can be found in Chapter 3.

6.6. Regression results and discussions

6.6.1. Preference for variety attributes using MNL model

The response variable in this choice model reflects preference among variety attributes: yield (Y), yield stability (YS), environmental adaptability (EA), marketability (M) and disease resistance (DR). When taking the most preferred attribute (in the MNL model), it should not imply that farmers are exclusively looking for a single variety attribute. Rather, the motivation is to study the factors that could induce farmers to have better preference or propensity for specific variety attributes. Tables 6-4 and 6-5 below report the MNL model estimation results.

Table 6-4. MNL regression results to explain farmers' demand for variety attributes

Variable	Coeff. for Y	Coeff. for YS	Coeff. for EA	Coeff. for M
Age	0.336 (1.03)	0.246 (0.78)	0.244 (0.79)	0.190 (0.55)
Agesqur	-0.0039(-1.17)	-0.0022(-0.69)	-0.0027(-0.85)	-0.0024(-0.67)
Educathh	-0.140(-0.98)	0.286^{**} (1.93)	0.217^{*} (1.45)	0.034 (0.22)
Anyincom	0.210 (0.20)	-0.269(-0.27)	0.110 (0.11)	0.179 (0.14)
Timarket	-0.0445^{*}(-1.72)	-0.030(-1.17)	-0.032^{*}(-1.26)	-0.058^{**}(-2.02)
Cofland	0.659 (0.30)	0.922 (0.40)	1.75 (0.74)	-0.530(-0.22)
Labor14	-0.561^{*}(-1.56)	-0.391(-1.09)	-0.556^{*}(-1.55)	-0.691^{**}(-1.90)
Land14	-0.6115(-1.74)	-0.486^{*}(-1.36)	-0.464^{*}(-1.32)	-0.440(-1.22)
$Risk_{proxy}$	0.0059^{***} (3.98)	-0.0032^{**}(-2.77)	-0.0028^{***} (-2.4)	0.0063^{***}(4.0)
Govrank	0.534^{***} (2.44)	0.0572 (0.28)	0.092 (0.46)	0.662^{***}(2.80)
Capital14	-0.1449(-0.41)	-0.0259(-0.07)	0.0366 (0.09)	-0.227(-0.60)
Naturank	0.6812^{*} (1.53)	0.620^{*} (1.45)	0.346 (0.76)	0.195 (0.39)
Harro	-4.19^{**}(-2.13)	-3.626^{***} (-2.05)	-3.98^{**}(-2.17)	-48.17^{***}(-21.8)
Kelaguda	21.48^{***} (21.89)	21.36^{***} (18.1)	21.57^{***} (18.16)	20.26
Kelokiri	1.045 (0.58)	1.727 (0.99)	1.23 (0.71)	0.853 (0.44)
Gibeboso	25.82^{***} (18.68)	22.22^{***} (13.5)	22.89^{***} (13.21)	24.61
Halosebe	22.97^{***} (23.93)	20.32^{***} (14.1)	21.10^{***} (15.03)	24.21
Sebekdeb	26.39	21.98^{***} (12.99)	22.89^{***} (13.87)	28.52^{***} (28.7)
Constant	-0.924 (-0.11)	-3.93(-0.46)	-1.74(-0.21)	3.37 (0.36)

Dependent variable is Attribute — Number of obs = 230

Pseudo R^2 = 0.4474 — Log likelihood = -188.23635

Disease resistance is left as a reference

Notes: ***-Significant at 1%; **- Significant at 5% and *- Significant at 10%. Two-tailed test is used for those variables the sign of the relationship is not predictable *a priori* and one-tailed test for those variables whose sign of relationship is predicted *a priori*. Figures in parentheses are the ratio of the coefficient to the estimated asymptotic standard error.

Table 6-5. Marginal effects of the MNL regression model

Variable	Dy/Dx: Y	Dy/Dx: YS	Dy/Dx: EA	Dy/Dx: M
Age	0.0148	-0.0079	-0.0069	-0.00001
Agesqur	-0.00024	0.00023	0.000017	5.3e-08
Educathh	-0.064	0.0493	0.01507	-0.00002
Anyincom	0.0505	-0.1024	0.052	0.000032
Timarket	-0.0022	0.0015	0.00068	-3.5e-06
Cofland	-0.103	-0.1057	0.209	-0.00025
Labor14	-0.016	0.0412	-0.0255	-0.00003
Land14	-0.022	0.0079	0.0141	9.4e-06
$Risk_{proxy}$	0.0015	-0.00088	-0.00057	1.09e-06
Govrank	0.0751	-0.0486	-0.0266	0.000073
Capita14	-0.0239	0.0010	0.02291	-0.00003
Naturank	0.0298	0.0373	-0.067	-0.00005
Harro*	-0.048	0.133	-0.020	-0.0651
Kelaguda*	0.0033	-0.0431	0.0399	-0.00012
Kelokiri	-0.070	0.139	-0.069	-0.00007
Gibeboso*	0.6713	-0.4163	-0.255	0.000079
Halosebe*	0.4763	-0.351	-0.126	0.0012
Sebekdeb*	0.752	-0.473	-0.283	0.0037

Note: (*) Dy/Dx is for discrete change of dummy variable from 0 to 1.

Over-all, the estimated MNL model is highly significant in explaining farmers' variety attribute preference. The Pseudo-R^2 was equal to 0.45. To see the predictive power of the model, the actual preferences and the ones predicted by the model were compared. The model correctly predicted about 73 percent of the yield preference, 82 percent of the yield stability preference, 96 percent of the environmental adaptability preference and 65 percent of the preference for marketability. The over-all prediction was 80 percent.

The results by and large show that factors inducing higher demand for yield (or marketability) are different from those factors leading to higher demand for yield stability (or environmental adaptability). For the most part, farmers' vulnerability to potential income shocks, market access, farm household concerns (input supply, extension and natural constraints), and the opportunity cost of resources (mainly land and labor) have been influential in determining variety attribute preferences.

The results suggest that farmers in more accessible areas and those who are less concerned with satisfying subsistence income prefer yield and marketability. On the contrary, farmers in less accessible areas and those more concerned with potential future income shocks (or survival) have more propensities for adaptability and yield stability.

As land and labor get more scarce, farmers' demand for yield increases. In contrast, a decrease in the opportunity cost of land and labor increases farmers' demand for yield stability. This could be because those farmers facing higher opportunity cost of land and labor or having better chance to get off-farm job opportunities (for labor) and cash crop farming (for their land) have a higher demand for yield to maximize income. This is against those who have no chance for their labor and land. These ones are maximizing survival through higher demand for yield stability and environmental adaptability. Farmers less constrained with land shortage tend to have higher demand for environmental adaptability. This could be because farmers who have large land holding (less constrained with land shortage) have more number of plots and heterogeneous land and thereby look for varieties that can be adapted to the different soil types. This is similar to previous research results which have found that farmers are trying to match different varieties to different soil types based on their suitability (Bellon and Taylor, 1993; Conelly and Chaiken, 2000). On the contrary, shortage of labor tends to increase farmers' demand for environmental adaptability.

Moreover, farmers who are less concerned with natural problems (disease, drought, and pests) have been found to have higher demand for yield. Farmers who expect good prospect for availability of rural development services (input supply, farm implements, markets, and extension services) have higher demand for yield and marketability. Thus, more productive assets like high yielding varieties will have better chance of acceptance by farmers provided that markets to absorb the surplus production accompany them.

All these results indicate the types of varieties that will be abandoned from the farm by specific types of farmers unless policy makers do not take counteractive conservation conservation measures. Similarly, the results show the types of varieties that will be of interest to specific types of farm households. Because of their effect on farmers' utility, rural development interventions in the areas of infrastructure development, poverty reduction, and market access will change the relative importance of the different variety attributes to households and thereby affect farmers' variety attribute preferences, demand for local varieties, and on-farm coffee diversity. For instance, if irrigation is made accessible to most small-scale farmers, they will be less concerned with attributes like drought resistance and environmental adaptability. Hence, those varieties embedding these attributes, which have been no more important, will be replaced by new varieties.

The results show that farmers less concerned with natural problems have higher demand for yield stability. This is seemingly contradictory to our expectation. However, it is mainly due to the positive relationship between disease resistance (the attribute left out as a reference) and yield stability. If disease resistance is over-riding constraint for farmers, their preference for yield stability, in comparison to disease resistance, will be lower. Concerning the mutual exclusive-ness of the options offered to farmers, one caveat should be noted before we close this discussion. Even though different in some respects, attributes like disease resistance, yield stability, and environmental adaptability are inter-connected and the choice options mayz not be mutually exclusive.

6.6.2. Farmers' demand for multiple coffee variety attributes

Both Poisson and ordered probit regression models are estimated to explain the attribute-based diversity index. While the ordered probit results are contained in Appendix IV-B (Table IV-8), Table 6-6 below reports the Poison regression results.

Table 6-6. Poisson regression results to explain farmers' aggregate demand for coffee variety attributes

Variable	Coefficient	Marginal effects: Dy/Dx
Experenc	-0.0020 (-0.89)	-0.0069
Educathh	0.00089 (0.10)	0.00306
Timarket	0.0033*** (3.35)	0.0114
Plots	0.0371* (1.74)	0.128
Purposes	0.141*** (4.92)	0.487
$Risk_{proxy}$	-0.00016*** (-3.79)	-0.00057
Landsize	0.0066 (1.19)	0.02286
Labendow	0.0396*** (3.11)	0.1365
Capital4	-0.0204 (-1.08)	-0.070
Govrank	0.0136 (0.82)	0.0467
Naturank	0.0024 (0.09)	0.0083
Yildrank	0.0786** (2.55)	0.271
Stabrank	-0.083*** (-3.45)	-0.286
Mrktrank	0.0586** (2.22)	0.202
Harro*	-0.359*** (-3.01)	-1.094
Kelaguda*	-0.240*** (-2.82)	-0.759
Kelokiri*	-0.1224 (-1.41)	-0.4026
Gibeboso*	-0.183 (-1.44)	-0.589
Halosebe*	0.167 (1.35)	0.611
Sebekdeb*	-0.018 (-0.15)	-0.061
Constant	0.452 (2.31)	
Dependent variable is *Locattribute*		Number of obs. = 242
Wald $\chi^2(20) = 289.60$		Prob > $\chi^2 = 0.00$
Log-likelihood = -433.9		Pseudo $R^2 = 0.118$

Notes: *** - Significant at 1%; **- Significant at 5% and *- Significant at 10%. Two-tailed test is used for those variables the sign of the relationship is not predictable *a priori* and one-tailed test for those variables whose sign of relationship is predicted *a priori*. Figures in parentheses are the ratio of the coefficient to the estimated asymptotic standard error. The method of estimation is Stata's Robust option following Huber/White standard errors and covariance. (*) Dy/Dx is for discrete change of dummy variable from 0 to 1.

Source: 2001/2002 original survey data

To examine farmers' effective demand for diverse coffee variety attributes, the number of attributes of coffee that each farmer has mentioned to be important for deciding on the use of local varieties was used as a response variable.

Predicted attribute diversity at the mean of the regressors is 3.45 which is very close to the average attribute count (3.69). Running Stata's 'Poisgof' option to perform goodness of fit test after Poisson results in a non-significant test *i.e.* goodness-of-fit χ^2 = 131.15 and Prob > χ^2 (221) = 1.00. This implies that the Poisson regression is appropriate (Statcorp, 2001).

The interpretation focuses on the change in the expected value of *Locattribute* as either of the independent variables change. The simple way to interpret the results is by using the factor change (Long, 1997). For instance, if the distance of the household residence from the market increases by 1 unit, the expected value of '*Locattribute*' increases by 0.33 percent and if the number of plots increases by 1 unit, the expected value of '*Locattribute*' increases by 3.71 percent. Coming to the dummy variables, being in Kelaguda decreases '*Locattribute*' by a factor of 0.76 (= exp [-0.24]) or decreases by 24 percent (0.76 - 1), holding all other variables constant.

The results show by and large that access to markets, number of plots, purposes of use of income from coffee, farmers' sensitivity to potential income shocks, and labor endowment of the household are the most important variables to explain farmers' aggregate demand for variety attributes.

The results further show that attribute count index increases with number of plots under cultivation, preference for yield stability, labor endowments, purposes of growing coffee, farmers' sensitivity to potential income shocks, and distance of the farmstead from markets.

The increase in number of plots increases farmers' aggregate demand for variety attributes because such farmers are endowed with heterogeneous plots that are suited to different types of varieties. Farmers who are capable of satisfying the labor intensity of multiple varieties have higher aggregate demand for diverse variety attributes. Both land heterogeneity and labor endowment increase the portfolio of variety attributes demanded by farmers because they have better incentives to utilize the resources they are endowed with. As the relative importance of coffee or number of purposes to which income from coffee is allocated increases, the diversity of attributes required by farmers also increases. This is a reflection of the fact that the relative importance of coffee to the household and the maximization of variety attribute portfolio are following the same pattern.

Farmers failing to satisfy their current household requirement have better incentive to diversify on their demand for multiple variety attributes. Risk averse farm households demand more attributes because, in the absence of certain prospects for better farm income, they can't depend entirely on a single variety attribute. Similarly, those farm households away from the markets have to satisfy their attribute preferences from within the household and hence the *attribute count index* is found to be higher for such farm households.

6.6.3. Farmers' effective demand for local varieties of coffee

With the motivation of understanding factors determining farmers' demand for local coffee varieties, both two-limit Tobit and OLS after logit transformation were estimated using the proportion of number of local coffee trees to total coffee trees on each farm as a response variable[10]. The results of these models are found to be pretty similar. The Tobit results are contained in Appendix IV-B (Table IV-9) and the OLS (after logit transformation) results are reported in Table 6-7 below.

[10]Proportion of coffee land allocated to traditional coffee trees was also attempted but due to the imprecision of farmers' estimates, the variables did not have adequate variation and could not be explained well using the regressors considered.

Table 6-7. Logit transformed OLS regression results to explain farmers' effective demand for local coffee varieties

Variable	Coefficient
Experenc	-0.021 (-1.02)
Educathh	-0.130* (-1.61)
Accesse2	0.025** (2.21)
Lanindic	-0.068 (-1.26)
Labindic	-0.092 (-0.31)
Regular	-4.01*** (-12.11)
$Risk_{proxy}$	-0.0012*** (-3.76)
Capital1	0.082 (0.62)
Govrank	0.169* (1.44)
Naturank	0.109 (0.64)
Beteryil	0.698** (2.08)
Betedrot	-0.373 (-1.29)
Betestab	0.418* (1.33)
Betemakt	-1.25** (-2.03)
Beterfod	-0.035 (-0.10)
Harro	1.56** (1.95)
Kelaguda	1.92** (2.07)
Kelokiri	-0.16 (-0.23)
Gibeboso	-2.13** (-2.25)
Halosebe	-3.43*** (-3.79)
Sebekdeb	-2.29*** (-2.67)
Constant	6.91 (4.08)
Dependent variable is logtree (log-transformed proptres)	
N = 240	F(21, 218) = 24.59
Prob > F = 0.00	R^2 = 0.5845
Adjusted R^2 = 0.412	

Notes: ***-Significant at 1%; **- Significant at 5% and *- Significant at 10%. Two-tailed test is used for those variables the sign of the relationship is not predictable *a priori* and one-tailed test for those variables whose sign of relationship is predicted *a priori.* Figures in parentheses are the ratio of the coefficient to the estimated asymptotic standard error.

After OLS, Stata's '*protab*' option was used to predict the effect of the significant variables on the proportions. The results suggest that the further away farmers are from markets, roads and extension, the higher the proportion of resources that they will allocate to FVs of coffee. As distance increases from 2 to 117, proportion increases from 0.92 to 0.995. This result is in line with the prediction of the adoption literature. For instance, for Malawian smallholders, Zeller and his co-workers (1998) have shown that with higher travel costs to the nearest market outlet, the share of hybrid maize declines.

Better-educated farm households have reported lower proportion of FVs of coffee trees on their plots. As the level of education increases from 0 to 13, the proportion declines from 0.97 to 0.86. Similarly, the higher the experience in the use of IVs, the lower the proportion of land allocated to local coffee trees. The effect of frequency of use of IVs is rather high. Moving from a state where no use of IVs at all to only a one time experience, proportion declines from 0.999 to 0.96. A further move to more frequent use leads to a collapse of the proportion to 0.29. Even though coffee farmers do not necessarily allocate marginal quality land to local coffee trees (see Sub-section 3.6.1), access to modern inputs are lowering their propensity to plant local coffee trees over the long-run.

The more vulnerable farmers are to potential income shocks, the higher the proportion of local coffee trees. As the gap moves from –2000 to +2500, the proportion drops from 0.99 to 0.33.

Over-all, local coffee varieties are mainly grown by less educated, less experienced (with IVs), those farmers found in less accessible areas (markets and extension) and subsistent farm households.

Unlike sorghum, the dying of the old generation and the emergence of the young does not disrupt the continuity of the local coffee varieties on farmers' plots. Many coffee growing farmers have reported to be growing local coffee trees that they inherit from their parents and grand parents.

Farmers whose major concern is lack of input supply, extension and markets have got lower proportion of local varieties of coffee on their plots. Thus, those farmers constrained by lack of public services (like input supply and extension) are not necessarily allocating the higher proportion of their plots to FVs. This does not contradict with the previous results because farmers' concern for access to extension is not a function of the availability of the extension service but its importance to household utility. A household could be located far away from the extension services but still not interested with it and may give lower rank to it. This lack of direct relationship between farmers' concerns with extension or markets and the distance of the farmstead from such facilities explains the seemingly contradictory result. As farmers' ranking of government services increases from 1 to 12, the proportion increases from 0.91 to 0.985.

Coming to the effect of farmers' preference among local and improved varieties of coffee, those farmers who have given better rank to the local varieties with respect to yield and yield stability have been found to have higher proportion of local coffee trees on-farm. Moving from households who gave better rank to IVs to those who gave better rank to FVs in terms of yield, the proportion increases from 0.94 to 0.985. On the contrary, farmers who have given better rank to the local varieties (with respect to marketability) have lower proportion of local coffee trees on-farm. As we move from households who gave better rank to IVs to those who gave better rank to FVs, the proportion declines from 0.988 to 0.87.

Thus, farmers' preference for yield and yield stability coincides with higher demand for local coffee trees while their preference for marketability coincides with lower demand for local varieties. The results imply that farmers whose preferred attributes are yield and yield stability allocate the majority of their land to local coffee trees while those whose preferred variety attribute is marketability allocate the majority of their coffee land to improved coffee trees.

According to the results, factors like farmers' experience in farming and the relative abundance of labor in relation to land do not play much of a role. Farmers' preference for local varieties with respect to food taste and drought tolerance has also no significant effect on the proportion of local coffee trees on-farm.

6.6.4. Farmers' effective demand for multiple local coffee varieties

Farmers' effective demand for multiple local varieties of coffee was analyzed using number of reported local varieties on-farm as a response variable and Poisson regression. Table 6-8 below contains the results.

Table 6-8. Poisson regression results to explain farmers' effective demand for count coffee diversity

Variable	Coefficient	Marginal effects: Dy/Dx
Experenc	0.0713^{**} (2.07)	0.012
Expsqur	-0.0013^{**} (-2.20)	-0.0002
Educathh	0.1135^{***} (3.46)	0.018
Anyincom	0.357^{*} (1.44)	0.058
Accesse2	-0.006 (-0.88)	-0.0009
Landsize	0.032^{*} (1.66)	0.0052
Plots	0.057 (0.69)	0.0093
$Labshort^{*}$	0.2038 (0.70)	0.036
$Risk_{proxy}$	-0.0004^{**} (-2.13)	-0.00006
$Capshort^{*}$	0.487^{**} (2.05)	0.073
$Govshort^{*}$	-0.328 (-1.38)	-0.050
$Natshort^{*}$	0.184 (0.65)	0.031
Yildrank	-0.39^{***} (-2.51)	-0.063
Stabrank	0.123^{*} (1.29)	0.020
Adaprank	-0.235^{**} (-1.94)	-0.0382
Mrktrank	-0.135 (-1.20)	-0.022
Diserank	-0.297^{***} (-2.54)	-0.048
$Harro^{*}$	1.655^{***} (2.66)	0.546
$Kelaguda^{*}$	1.606^{***} (2.40)	0.524
$Kelokiri^{*}$	1.280^{*} (1.78)	0.364
$Gibeboso^{*}$	3.564^{***} (5.46)	3.79
$Halosebe^{*}$	2.822^{***} (4.44)	1.73
$Sebekdeb^{*}$	2.651^{***} (3.79)	1.47
Constant	-3.59 (-3.43)	

Dependent variable is localvar	Number of observations = 249
Wald $\chi^2(23)$ = 107.42	Prob > χ^2 = 0.00
Pseudo R^2 = 0.1946	Log likelihood = -132.3

Notes: ***-Significant at 1%, **- Significant at 5%, and *- Significant at 10%. Two-tailed test is used for those variables the sign of the relationship is not predictable *a priori* and one-tailed test for those variables whose sign of relationship is predicted *a priori*. Figures in parentheses are the ratio of the coefficient to the estimated asymptotic standard error. The method of estimation is Stata's Robust option following Huber/White standard errors and covariance. (*) Dy/Dx is for discrete change of dummy variable from 0 to 1.

Running stata's 'Poisgof' option to perform goodness of fit test after Poisson results in a non-significant test (Goodness-of-fit χ^2 = 130.9 and Prob > χ^2 (225) = 1.00) implying that the Poisson regression is appropriate.

The results reveal that as the vulnerability of farmers to potential income shocks increases, the number of local varieties of coffee maintained increases. Those farmers whose major concern is yield, disease resistance, and adaptability to environmental stress have maintained more local varieties on-farm.

As farmers' experience in farming increases, the number of local coffee varieties maintained on-farm also increases. Once farmers gain enough experience with the available local varieties, the variety count on-farm then declines. More educated farmers also maintain more number of local varieties. Not only education level of the household head but also income outside agriculture has positive effect. While working capital shortage affects variety count positively, access to input supply, extension, and farm implements as production constraints was not significant.

Unlike sorghum, distance from markets, extension and roads is not important for explaining coffee count diversity. While land size holding has positive impact on count coffee diversity, the number of plots were found to be not so important.

6.7. Summary and concluding remarks

This chapter has explained farmers' preferences for variety attributes, demand for local varieties of coffee, demand for variety attribute count, and demand for variety count using a variety of econometric models. For the most part, farmers' vulnerability to potential income shocks, market access, farm household concerns (input supply, extension and natural constraints), experience in growing IVs, and the opportunity cost of resources (mainly land and labor) have been influential in determining these variables. Overall, local coffee varieties are mainly grown by less educated, less experienced (with IVs), those farmers found in less accessible areas (markets and extension), and subsistent farm households.

Because of their effect on farmers' utility, rural development interventions in the areas of infrastructure development, poverty reduction, irrigation, and market access will change the relative importance of the different variety attributes to households and thereby affect farmers' variety attribute preferences, demand for local varieties, and on-farm coffee diversity. For instance, if irrigation is made accessible to most small scale farmers, they will be less concerned with attributes like drought resistance and environmental adaptability. Hence, those varieties having these attributes will be replaced by new ones.

Targeting is necessary not only for on-farm conservation but also for variety adoption. For instance, policy can target the more vulnerable farmers for on-farm conservation due to their pro-diversity decision behavior. The details of the conclusions and policy implications of the empirical results are contained in Chapter 9.

7. Opportunity cost of maintaining local varieties of sorghum and wheat on farmers' fields

7.1. General background

Policy makers have to be aware of the fact that farmers can not afford to maintain local varieties on their farms just for the sake of crop diversity. From the choice set available to them, farmers are planting indigenous varieties of their respective interest. There could still be other varieties not of interest for any farmer. Before designing incentive mechanisms to induce farmers to plant these varieties, decision-makers have to be informed on the farmers' expectations (in terms of income or variety attribute losses). This can be derived from opportunity costs farmers are facing when they plant varieties that need to be maintained for policy purpose.

Selection of farmers and sites for on-farm conservation purpose is so far done without detailed investigation of contextual factors affecting farmers' opportunity costs. Ethiopia's Biodiversity Institute has recently been engaged in on-farm conservation of landraces of different crops in different parts of the country. For instance, the GEF financed project has started the Ethiopian on-farm landrace conservation and enhancement program involving the IBCR, breeders, scientists, and farmers. Agreements were signed with farmers to the effect that they would conserve specific crop varieties and the IBCR would compensate them for the yield loss compared with the yield from IVs (Demissie and Arega, 2000). Yield of the crops being conserved by the farmer conservators was evaluated and compensation for yield difference between improved and local varieties was determined. Compensation was then paid based on this difference. This approach, however, has many problems.

The IBCR was using a contract that depends on the outcome. If a farmer gets higher yield from the varieties maintained on-farm, he/she will get lower compensation. This clearly creates a *moral hazard* problem because farmers will not have incentive to productively use their varieties since they will get less compensation if they get good yield from the varieties maintained. How can policy tackle such a problem? To handle this problem, this chapter argues that the level of compensation should be decided *ex ante* considering the opportunity costs faced by representative farm household types. Moreover, considering only yield differences does not give full picture of opportunity costs as IVs are usually used along with modern inputs like fertilizer and recommended practices. Payments based on yield differentials will, therefore, attract many farmers to the scheme that makes targeting difficult and susceptible to corrupt practices.

Many incentive schemes are designed based on cost data (Laffont and Tirole, 1993). In this study, the cost information is based on what happens to the income of the farm household when on-farm conservation is implemented *i.e.* the farmer is expected to plant FVs worth maintaining on-farm. This computation is made based on the sorghum and wheat household survey data described in Chapter 3.

The starting point to compute opportunity cost is yield differences between local and improved varieties. How ever, this difference does not fully depict the opportunity cost story. Higher yield from IVs could be achieved using higher input levels. To depict the full picture of opportunity costs, there is a need to put both yield and input levels in monetary terms. To this end, the gross margins for the respective varieties (IVs and FVs) are computed for each crop.

As far as monetary value of varieties is concerned, the gross margin (here after GM) from the IVs is the first best outcome from use of sorghum or (wheat) farm land. This is assuming that, on the whole, IVs are more rewarding than FVs. Secondly, the GM difference gives the order of magnitude for on-farm conservation costs in economic terms as it is the amount of money that the country is foregoing for the sake of on-farm conservation.

The remaining part of this chapter is structured as follows. Section 7.2 provides the reasons for dealing with opportunity costs and presents the research objectives. This is followed by the theory in Section 7.3. The specifications of the econometric models are presented in Section 7.4. Section 7.5 takes up a brief description of the variables used in the regression. Estimation results are presented and discussed in Section 7.6. Finally, Section 7.7 summarizes the chapter and provides some concluding remarks.

7.2. Why opportunity costs matter for on-farm conservation?

If society wants to have more diversity than what the *status quo* produces, direct and/or opportunity costs have to be incurred. One of the means to create incentives for farmers to participate in on-farm conservation ventures is to compensate what they are losing when they are expected to change their variety choice behavior for the sake of maintaining specific variety (ies). The issue that policy faces is how to induce farmers to comply with the optimal level of effort set (number and types of FVs to be maintained on-farm) to achieve the on-farm conservation objectives. What is not known yet regarding opportunity costs is 'How high should the compensation be and on what factors does it depend?' The analysis done in this chapter is meant to contribute to this question and show what farmers can expect from policy when on-farm conservation is put in place.

In any attempt to influence farmers' variety choice behavior, policy needs to ask the question: 'What is it that farmers will have to sacrifice if there is a need to change their variety choice behavior for the sake of maintaining crop diversity on-farm?' Addressing this question appropriately enables designing appropriate compensation scheme based on what farmers are losing when policy wants them to change their variety choice. Any loss (monetary or non-monetary) faced by farmers when they switch from varieties under use to varieties worth maintaining is what is called the opportunity cost of changing variety use. The financial opportunity cost is:

$$OPPORTUNITY\ COST = (GMPH_{IVs} - GMPH_{FVs}) \qquad (7.1)$$

where $GMPH_{IVs}$ and $GMPH_{FVs}$ refer to GM per hectare (in Birr) of the improved and local varieties, respectively[11]. Computing opportunity costs can have important role in the incentive design because the bargaining power of the farm households is mainly a function of the opportunity costs they face. GM does not account for opportunity costs of fixed factors of production like capital. If capital is a binding constraint and IVs use more unit of capital than FVs, the above computation over-estimates opportunity costs of growing FVs. However, all financial costs involved in producing both varieties have been included in the GM computation.

Having decided on the varieties to be maintained and the level effort expected by farmers, policy makers have to compensate them the financial and non-financial loss incurred consequently. In computing the financial opportunity costs, no attempt is made to integrate the non-monetary benefits of using multiple FVs (like food tastes, cooking quality, livestock feed, risk reduction, and early maturity) in the calculation. This can lead to under or over estimation of the opportunity costs depending on the attributes

[11]In computing opportunity costs this way, we are assuming that the first best alternative use of a wheat or sorghum plot is IVs of the respective crops.

embedded in specific varieties and thereby lower (higher) the compensation. Consequently, the targeted farmers will get less (more) than what they should. However, the varieties being used by farmers may not be all the time superior or inferior to what policy wants them to use and maintain. In Section 3.6, it has been shown that neither IVs nor FVs are superior with respect to all variety attributes. Thus, our failure to integrate the non-monetary benefits is expected to have no serious consequence on the opportunity cost calculation.

Each set of FVs used in each locality has different desirable features *i.e.* FVs being used by farmers in village A may be superior in food tastes but those in village B could be superior in early maturity. Farmers' concerns and constraints in different localities also differ *i.e.* drought tolerance could be the most important concern in village A but early maturity could be the priority concern in village B. To address this issue, the variety attribute trade-offs between what the farmers are planting and what they are supposed to grow for policy purpose needs to be understood contextually.

Despite all these benefits of including variety attributes in the opportunity cost calculation, it is very hard to get the monetary value of the compensation required to recompense the non-monetary trade-off faced by the targeted farmer(s). Moreover, the equivalence of different variety attributes is not known *i.e.* how many times is disease resistance more important than food quality or the *vice verse*. In this chapter, no attempt is made to convert the non-financial attributes to monetary terms. The chapter rather concentrates only on the financial level of compensation required based on the GM comparison from improved and local varieties. The results of Chapter 3 on comparison of the IVs and FVs with respect to different variety attributes and Chapter 6 on farmers' variety attribute preferences should, however, complement the information to be used in the design of incentive schemes as is shown in Chapters 8 and 9.

As far as compensation is concerned, the most important question that needs contextual research is: 'How high should the financial compensation be and what factors determine the amount of compensation required?'[12] If it is the

[12]The idea of compensating farmers may be politically controversial in the policy circle. However, if the government is prepared to pay the opportunity cost of conserving crop diversity at the national level, there is no reason why targeted farmers should not be compensated for their contribution based on what they are sacrificing. Expecting farmers to maintain varieties of policy interest without compensating them the loss is forcing them to cooperate with policy without their will. This, clearly, does not work.

political will of the Ethiopian government to design incentives that are compatible with farmers' expectations, studying opportunity costs will inform policy makers on the required levels of compensation. Considering the GM from the IVs as the upper bound for compensation, a question worth asking could be: 'What is the extra GM that users of IVs (hereafter UIVs) gain compared to users of FVs (hereafter UFVs)?'.

Having these points in mind, the research objectives of this chapter include:

- quantifying the opportunity costs of maintaining local varieties of sorghum and wheat on-farm;
- analyzing the contextual factors affecting opportunity costs of maintaining local varieties on farmers' fields; and
- generating information on the level and types of incentives required for on-farm conservation of local varieties in Ethiopia.

Financial opportunity cost reflects the amount of money that farmers are foregoing to enjoy the non-yield benefits of FVs. From this perspective, the opportunity cost could be considered as farmers' willingness to pay (in terms of GM foregone) for the sake of enjoying desirable attributes of local varieties. The point of interest for this chapter can thus be rephrased as 'How does this willingness vary among households of different features?'

Wherever the conservation of biodiversity does have an opportunity cost in terms of output, the aggregate value of output foregone provides a first approximation of the value to the conserving authority of the maintenance of ecosystem resilience through conservation (Perrings, 1995). Thus, the aggregate values of opportunity costs we are computing can be considered as the gross value of crop diversity to the national economy.

Despite its immense role in national conservation policy, studies on the opportunity cost of crop diversity conservation are non-existent. As far as our knowledge is concerned, the only study available so far is the aggregate level estimate of opportunity cost of biodiversity conservation in Kenya by Norton-Griffiths and Southey (1995). This chapter aims to contribute to this research gap at the household level.

7.3. Theoretical framework

7.3.1. Issues in the opportunity cost definition and factors affecting it

As shown in Chapter 6, given their choice set, farmers' variety choice is a question of the fitness of attributes of variety (ies) to their conditions and household concerns. However, choice and use of any variety, be it local or improved, involves trade-offs and opportunity costs. While choosing certain variety (ies) from the choice set to meet certain objective(s), a farmer loses other important attributes from the set of varieties not selected.

From the utilitarian vantage point of view, to be maintained on-farm, each additional variety (be it local or improved) must have a positive marginal utility to the farm household which the other variety (ies) in use can't offer. According to the random utility theory, a farmer uses IVs if the utility of adopting the technology is at least as large as the utility without its use. If farmers are using multiple local varieties for the sake of reducing risk by foregoing a positive extra net financial benefit from the IVs, the financial benefit foregone can be considered as a risk premium. The higher the level of the risk premium, the higher will be the opportunity cost.

One of the forces driving farmers to maintain or replace varieties on-farm is the opportunity cost foregone because farmers' propensity to use a certain variety is, among other things, dependent on the expected benefit from growing other varieties from their choice set. If the opportunity cost of growing local varieties is increasing in a given locality, the farmers facing higher opportunity cost should be allocating more proportion of their resources to IVs[13].

Having the definition of opportunity cost (Section 7.2), its size depends on the suitability of farmers' environment to the production and marketing of improved and local varieties. The factors determining the suitability include access variables (access to markets and extension), level and quality of input use (fertilizer, chemicals, oxen, and land), input and output price levels, resource endowment (education, labor, and cash crop farming), experience in growing IVs, and natural factors (rainfall, drought, disease, pests, *etc.*). The more favorable these conditions are to the production and marketing of the IVs compared to FVs, the higher will be the size of the opportunity cost to be foregone. Inputs and local conditions affecting GMs from both

[13]The farmer only has opportunity cost when the utility (in terms of GM) of adopting the new variety is larger than the utility of the local variety.

varieties equally do not affect opportunity costs. Drawing from this theoretical sketch, the predicted effects of the respective variables on the opportunity costs are presented in Tables 7-1 and 7-2 below.

The lower the predicted probability of the survival of FVs, the higher will be farmers' opportunity costs (Smale, 2003)[14]. Thus, the in-built incentives (examined in Chapter 5) are in theory inversely related with opportunity costs. The specific hypotheses to be tested in this chapter include:

- The higher the market price per unit for the crop, the higher the opportunity cost[15] *i.e.* opportunity costs of wheat are higher than that of sorghum;
- The opportunity cost of maintaining local varieties on-farm is an increasing function of better access to input and output markets; and
- The more experienced farmers are with IVs, the higher will be the opportunity costs involved.

7.3.2. Theory behind the choice of methods of data analysis

Of course, the simple GM difference (as presented in Section 7.2) is not the result only of use of IVs. There are also other household and environment related factors that affect the GM outcomes in the two regimes. These factors are not randomly distributed among UIVs and UFVs. The simple definition given in Section 7.2 will hold true only if all farmers are using both varieties (improved and local) side by side on the same quality plots. In that case, all the household and environment related differences across farmers will be controlled.

Due to non-random distribution of the non-variety factors (factors affecting GM of IVs and FVs) and unobserved variables, *selection bias* could arise in the empirical analysis. Sample selection bias may arise in practice for two reasons (Heckman, 1979). First, there may be *self-selection* by the farm households themselves and secondly sample selection decisions may be made by analysts or data processors. Both problems arise because the variety

[14]Those farmers with less interest with FVs and still planting them for other technical reasons are, by *de facto*, paying higher opportunity costs because they would have benefited more from IVs.

[15]This is because the higher the price of a crop per unit, the higher will be the monetary value of the foregone benefit.

choice decisions will no more be random and that non-random decision process is affecting GM from IVs and FVs differently.

In the present case, the first version of the *self-selction* problem is due to farmers' own self-selection in such a way that each farmer takes the variety (ies) that have comparative advantage to his/her working environment. Under *self-selection*, farm households who have a better comparative advantage with the use of IVs will join the use of IVs and thus will benefit more from it than would be the case for a randomly selected farmer. The second type of self-selection could arise due to the differential treatment of the Ethiopian government (Agricultural Extension Department, MoA) targeting the so called 'high potential areas or farmers' in the distribution of IVs and accompanying inputs. If the UIVs have better access to market, credit, extension, education, irrigation, and other productivity promoting factors, this could make them more productive than a randomly selected farmer *i.e.* UIVs would earn higher GM even from their local varieties. If the *sample selectivity bias* problem is not taken care of, the value of IVs will be potentially over-estimated.

Selection-bias could be the result of selection on the observables or selection on the unobservables. Better educated farmers, better quality land, and use of better productive inputs could be skewed towards the UIVs. This is selection on the observables. The GMs achieved by farmers using improved or local varieties are, accordingly, affected differently by the explanatory variables. Selection on the observables can be controlled by having all the variables in the GM equations of the response variables (Vella, 1998). Regarding selection on the unobservables, the essence of the problem is that UIVs and UFVs are not the same with respect to variables relegated to the error term. Given that GMs for UIVs and UFVs are observed conditional on different unobservable factors, there will be a *self-selection* problem in the observed GM data (Huang, Raunikar and Misra, 1991). One most important variable that is hard to capture, for instance, is farmers' managerial capacity. If unobservable variables are correlated with the observables, our inference about the impact of the observables on GM is biased (Vella, 1998) due to selection bias.

To sum up, selection bias arises when the mean GM of the UIVs differ from the GM of comparison group farmers even in the absence of improved variety use (Heckman and Hotz, 1989). Selection bias does not arise if selection depends exclusively upon the exogenous variables only (Vella,

2000). Disregarding self-selection, a GM equation that takes into account the value of using IVs as a dummy variable could be set using the equation:

$$GM_i = \beta' x_i + \delta imp_i + \varepsilon_i \tag{7.2}$$

where GM_i is the GM for the ith farmer, x_i's are the independent variables, and Imp_i is a dummy variable indicating whether or not the farmer has used improved variety. However, the coefficient δ does not measure the value of improved variety if the typical farmer who chooses to use improved variety would have relatively high GM irrespective of improved variety use (Greene, 2000, p. 933). Of course, OLS estimate of δ will over-estimate or under-estimate the treatment effect depending on the type (positive or negative) of self-selection. The simple OLS regression taking use/non-use as a dummy is implying that the two groups of farmers are drawn from a random sample or they have the same potential to earn GM irrespective of their variety choice. Due to non-random distribution of non-variety factors, this has not been the case in the data-sets employed here. The econometric model employed here is meant to address this issue.

The other option one can think of is to estimate separate OLS equations for UIVs and UFVs. This does not solve the problem either because the two equations are not comparable due to the self-selection problem just described.

Splitting each data-set into two, a simple Chow test was run to test whether coefficients differ by variety use status. In both data-sets, the test rejects the hypothesis that the two regressions are the same. This supports the idea of using a regression approach that distinguishes each coefficient for the two groups. Had the Chow test resulted in insignificance of the difference, a simple OLS (instrumenting variety use dummy) could have been enough on the pooled data (Goldfeld and Quandt, 1973). For this reason, a switching regression model is used to study the factors affecting opportunity costs[16]. Heckman's two-step selection model was estimated to see the prevalence of the problem. The results are reported in Appendix IV-C (Tables IV-12 and IV-13). The significance of the inverse mills ratio suggests the prevalence of the problem of selection bias. Having justified the kind of econometric model worth considering, the section that follows presents the empirical methods of data analysis adopted.

[16]A switching regression model can be used to correct for possible *selection bias* problem (Freeman *et. al*, 1998).

7.4. Econometric model specifications

Once the GM per hectare for both groups of farmers is computed, the next step is to see as to how the opportunity cost varies along different farm households and localities considering different variables. In essence, this chapter is an evaluation study of the effect of improved variety use on GM per hectare because comparing UIVs and UFVs with respect to GMs gives it a treatment-control flavor. The most important problem in any evaluation study is the lack of a counterfactual *i.e.* it is not possible to observe what would the GM of UIVs have been had they been UFVs and the *vice verse* (Blundell and Costa-Dias, 2000).

If analysts could observe improved and local variety use GMs (Y_1 and Y_0) for the same farmer, there would be no evaluation problem (Heckman *et.al*, 1997; Heckman and Robb, 1985). In that case, one could form the conventional measure of gain or loss: $\Delta Y = Y_1 - Y_0$. Even though the problem is exceptionally acknowledged and discussed in the literature, no evaluation methodology so far guarantees a complete solution to the missing data problem (Blundell *et.al*, 2001; Rossi *et.al*, 1999). Blundell and Costa-Dias (2000) have reviewed different evaluation methods in empirical economics. For the purpose at hand, homogeneous treatment effects statistical models (matching and treatment effect regression) will be used to estimate the average opportunity costs and the switching regression model is employed to study the effect of different contextual factors on the opportunity costs. While computing the average opportunity costs is supposed to give indication on the levels of compensation, analysis on the factors affecting the opportunity costs can inform policy on the conditions under which lower (higher) payments are called for.

7.4.1. Homogeneous treatment effect models

In homogeneous treatment effects models, it is assumed that every farmer benefits equally from improved variety use. Even though it is hard to defend this assumption, these models can generate information that provides an indication on the level of compensation required. In this chapter, different methods of generating the average opportunity costs are used to see the sensitivity of the results to the choice of the specific method.

Matching

Matching is an evaluation method based on the intuitively appealing idea of contrasting the GM outcomes achieved by the UIVs (denoted y_{1i}) with the

GM outcomes of 'comparable' UFVs (denoted y_{0i}) so that the differences ($\Delta y_i = y_{1i} - y_{0i}$) in outcomes between the two groups can be attributed to the use of IVs. In this sense, matching is selection on the observables *i.e.* it involves selecting UIVs and UFVs based on their observable characteristics so that they are the same except by the type of variety used. It is an effort to draw a 'partner' for each UIV from the group of UFVs. By putting UIVs and UFVs in a similar footing except in variety use and making the respective farmers similar except in the use of improved varieties, it tries to minimize the bias.

Matching estimates the propensity score (the probability of belonging to group of UIVs given observable characteristics) in step 1 (using logit or probit) and uses the predicted value of the first step estimation for finding a counterpart for each UIVs from among those farmers using local varieties (Rosenbaum and Rubin, 1983). In the second step, matching pairs to each UIV one or more 'comparable' UFV (in terms of the matching variable) and associates to the outcome of the UIV the (weighted) outcomes of his 'neighbors' in the comparison group where the weights can optionally depend on their 'distance' to the UIVs under consideration. Matching has been extensively used in the evaluation literature (Heckman *et.al*, 1997; Heckman *et.al*, 1998).

From the matching exercise, the average treatment effect of using the IVs could be either negative or positive. If the over-all effect is positive, it implies that there is an over-all comparative advantage for producing IVs. But still from among those farmers included in the analysis, there could be farmers who have no comparative advantage in using the IVs.

Treatment regression model

On the premises that using IVs has merely an intercept effect on GM (*i.e.* results in a parallel shift up or down), the appropriate model instruments improved variety use status as a right hand side dummy. In instrumenting, all exogenous variables in the system must be put as instruments for efficiency reason (Baltagi, 1999). Before running 2SLS, the endogeneity of two suspected variables ('*user*' and '*impexep*' in the sorghum data-set and '*improvex*' and '*exteexp*' in the wheat data-set) have been tested using the Hausman test. The F- and Wald test results have rejected combined endogeneity of both variables in the respective equations whereas checking the two variables independently has led us to accept the hypothesis of endogeneity for the variables '*user*' and '*improvex*'. Hence, these two variables have been instrumented in the respective equations.

The treatment regression in Stata (version 7.0) considers the improved variety use dummy as dichotomous by fitting a probit equation model. The reason for using treatment regression is that the random shocks that affect a farmer's GM also affect whether that farmer uses IVs or not.

The treatment regression function takes the form:

$$y_j = x_j\beta + \delta z_j + \varepsilon_j \tag{7.3}$$

where z_j is an endogenous dummy variable indicating whether IVs are used or not. The binary decision to obtain the treatment z_j is modeled as the outcome of an unobserved latent variable, Z_j^*. It is assumed that Z_j^* is a linear function of the exogenous covariates w_j and a random component u_j. Specifically,

$$z_j^* = w_j\gamma + u_j \tag{7.4}$$

and the observed decision is: $z_j = \begin{cases} 1, & if\ z_j^* > 0 \\ 0, & otherwise \end{cases}$

where ε_j and u_j are bivariate normal with mean 0 and covariance matrix $\begin{bmatrix} \sigma & \rho \\ \rho & 1 \end{bmatrix}$.

Accordingly, the average difference in expected earnings between UIVs and UFVs is:

$$\text{Opportunity cost} = E(y_i \mid z_i = 1) - E(y_i \mid z_i = 0) = \delta + \rho\sigma\left[\frac{\phi(w_i\gamma)}{\Phi(w_i\gamma)\{1 - \Phi w_i\gamma\}}\right] \tag{7.5}$$

If the correlation between the error terms (ε_j and u_j), ρ, is zero (which is not the case in both data-sets), then the problem reduces to one estimable by OLS. This is because when the two error terms are not correlated, the difference will simply be δ. Since ρ is found to be negative in our case, OLS understates the treatment effect, δ[17] (see Table 7-3 in Section 7.6 and Tables IV-10 and IV-11 in Appendix IV-C). The net treatment effect depends on the relative size of δ and the second term that could lead to over-all negative or positive effect. According to this estimation, the coefficients of the x's are restricted to be the same for UIVs and UFVs. For instance, the return to

[17]The under-estimation is equal to the second term in the right hand side of equation 7.5.

fertilizer is restricted to be the same regardless of whether a farmer is in a UIVs group or not. The model in Sub-section 7.4.2 will handle this issue.

The statistical procedures above (matching and treatment effects regression) focus only on estimating mean impacts of using IVs. However, the more interesting question could be 'Who benefits most from IVs?' or 'What factors determine the size of the opportunity cost?' Addressing this question requires estimating GM equations for UIVs and UFVs and comparing the coefficients of the respective variables in the two equations. This is done here using the switching regression model presented below.

7.4.2. The switching regression model

If one believes that improved variety use does not only have an intercept effect but also a slope effect (*i.e.* the estimated coefficients differ according to variety use status as well), then a switching regression is the appropriate model to be used (Goldfeld and Quandt, 1973; Quandt, 1988). According to this model, the effect of using IVs does not show up as a dummy variable but rather in the fact that the constant term and other coefficients may differ across UIVs and UFVs sample. Essentially, this model allows full set of interaction terms between variety use status and the explanatory variables.

Let us consider the usual linear regression setting:

$$y_i = x_i \beta_i + e_i \qquad (7.6)$$

Taking this basic equation, it is possible to split it into two regimes that generate our dependent variable (GM per hectare from sorghum and wheat farming) as (Maddala, 1983, page 261):

$$y_{1i} = \sum_{j=1}^{k} \beta_{1j} X_{ji} + u_{1i} \quad \text{(Regime 1 which holds if } C = 1\text{)} \qquad (7.7)$$

$$y_{0i} = \sum_{j=1}^{k} \beta_{0j} X_{ji} + u_{0i} \quad \text{(Regime 0 which holds if } C = 0\text{)} \qquad (7.8)$$

$$C^* = \gamma_j Z_{ji} + u_i \qquad (7.9)$$

$$C = 1 \;\; if \;\; C^* > 0 \Rightarrow C = 1 \; if \; u_i > -\gamma_j Z_{ji}$$
$$C = 0 \;\; if \;\; C^* \le 0 \Rightarrow C = 0 \; if \; u_i \le -\gamma_j Z_{ji}$$

where the errors, u_{1i} and u_{0i}, are assumed to be distributed normally and independently with mean zero and a constant variance (σ^2). The γ_j are unknown coefficients to be estimated and Z_{ji}'s are latent variables that determine from which regime the ith observation is generated. The X_{ji} refer to factors affecting GM in the respective regimes (See Tables 7-1 and 7-2 for the list of variables). This model implies that the unobserved disturbances (u_i) of the selection criterion equation are correlated with the unobserved disturbance of u_{1i} and u_{0i} of the GM equations. It is simultaneously modeling choice between local and improved varieties and gross margins of both options. The point of interest is the correlation between the factors affecting variety use decision and gross margin in both regimes.

If C is the function that determines the switching from one regime to the other (the so called *criterion or choice function*), y_{1i} and y_{0i} are the improved and local variety GM outcomes, and μ_{1i} and μ_{0i} are the expected values of y_{1i} and y_{0i}, respectively, the two regimes can be combined as:

$$y_i = C(y_{1i}) + (1-C)y_{0i} \quad = C(\mu_{1i} + u_{1i}) + (1-C)[\mu_{0i} + u_{0i}] \tag{7.10}$$

This is what is referred to as the *switching regression* model (Quandt, 1988) that switches between regimes y_{1i} and y_{0i}, respectively, depending on the value of the criterion function *i.e.*

$$y_{1i} = \mu_{1i} + u_{1i} \quad \text{if } C \text{ equals } 1$$

$$y_{0i} = \mu_{0i} + u_{0i} \quad \text{if } C \text{ equals } 0$$

The model switches between y_{1i} and y_{0i} according to a dichotomous 'switch' (C^*) that itself is explained (Quandt, 1988). This is the so called the mover-stayer model. The observed y_i takes the value y_{1i} if C equals 1 and y_i takes the value y_{0i} if C equals 0. Assuming that the mean values of u_{1i} and u_{0i} are zero, the expected values of the outcomes in the two regimes and the expected values of the error terms can be presented as:

$$E(y_{1i} \mid x_i) = \mu_{1i} \quad \text{and } E(y_{0i} \mid x_i) = \mu_{0i} \tag{7.11}$$

$$E(u_{0i} \mid x_i, C=1) = E(u_{1i} \mid x_i, C=1)$$

The last term is the mean selection bias (Heckman, 1997).

The switching regression can involve known sample separation or unknown sample separation. In case where the sample separation is known, the researcher knows in advance to which regime each observed GM belongs to. Other wise, the researcher does not have information on whether a given GM is from regime 1 or regime 0. Presently, the GMs are observed in both regimes with known sample separation that endogeneously determines the regime. This kind of model is what is called switching regression model with endogeneous switching (Madalla, 1983). A full information maximum likelihood estimator that jointly estimates the coefficient and variance parameters from LIMDEP is implemented in Tables IV-12 and IV-13 of Appendix 4-C.

To interpret the coefficients of the switching regression model for the purpose of analyzing opportunity costs, the relative size of the respective coefficients in the two regimes have to be compared. After getting consistent estimates for these parameters using full information maximum likelihood (FIML), it is possible to calculate estimates of y_{1i} and y_{0i}. The size and sign of the GM difference in the two regimes ($\hat{y}_{1i}$ - $\hat{y}_{0i}$) is the point of interest in the analysis as an indicator for the opportunity cost.

If this difference is zero, it implies that users of FVs are indifferent in their choice of either IVs or FVs. This, in turn, implies that neither IVs nor FVs have comparative advantage. If it happens to be greater than zero, it implies that non-users of IVs face opportunity costs. This, in turn, demonstrates that there is comparative advantage to use IVs. If it happens to be less than zero, it implies that UIVs face opportunity costs. This is an indicator for the comparative advantage of FVs.

Estimating the switching regression model enables us to estimate GM for each farmer in the two regimes. The predicted difference is the opportunity cost of non-use of IVs. To see how opportunity costs are affected by different regressors, the coefficient of each regressor in regime 1 is subtracted from its respective value in regime 0. If this value is positive, the opportunity cost is affected positively and if it is negative, the opportunity cost decreases with that specific variable. Not only the effect of specific variables on opportunity costs but also the over-all impact of IVs on GM can be evaluated using switching regression model.

One of the potential uses of switching regression models is to evaluate the benefits of social programs (Maddala, 1983). To this end, the expected GM of UIVs and the expected GM of UFVs had they used IVs have to be compared. As it is not the purpose of this study, evaluating the over-all impact of using IVs is not pursued here.

7.5. Description of variables

The data generation process is described in Chapter 3. The relative importance of different variety attributes and farmers' evaluation of the improved and local varieties are also presented there. This exercise has revealed that yield was one of the most important variety attributes for farmers. Most of the other variety attributes can also be translated into yield. As it is shown in Section 3.6, the IVs of sorghum get favorable evaluation with respect to yield, disease resistance, and drought tolerance. Thus, designing the incentive schemes based mainly on the GM difference between the IVs and FVs has reasonable potential to be accepted by farmers in addition to simplifying the implementation of the incentive design.

To study the effect of input and output price levels on the opportunity cost, input and output price indices are computed. Output price index is computed using the conventional way as the ratio of the price of output that the ith household faces to the over-all average price. The average price is taken as the counterpart of the base year price in the case of time series data. To construct the input price index, for n inputs used in producing sorghum and wheat, the *weighted input price index* is computed in two steps. First, the individual input price indices (ψ_{ij}) are computed for each household using the same procedure as the case of the output price index. Adding all these price indices to get the over-all input price index could seem plausible but the relative importance of each input differs across households. For this reason, the ratios of the ith input cost to total cost (η_{ij}) are calculated for each household using market and imputed values of all inputs to which price indices are computed. These ratios are used as weights in the computation of the input price indices for each household. The ratio tells the contribution of the ith input in the total cost structure of the crop to produce sorghum and wheat. Thus, the input price indices (κ_{ij}) are computed as: $\kappa_{ij} = \sum_{i=1}^{n} \psi_{ij}\eta_{ij}$ where i indexes households and j indexes inputs.

The GM of the improved and local varieties is computed on a per capita basis as the gross revenue from the respective crops minus costs of variable inputs (seeds, fertilizer, labor used for different cultural practices – land preparation, sowing, cultivation, weeding and harvesting, cost of bullock labor, cost of chemicals, and other costs such as labor costs incurred for bird watching).

Tables 7-1 and 7-2 below report descriptive statistics for the variables used later in the regression.

Table 7-1. Descriptive statistics of the variables used in the opportunity cost regression: sorghum

Variable	Description	Mean (SD) – UIVs[18]	Mean (SD) – UFVs	Expected effect on OCs
GMM	GM per hectare (Birr)	427.6 (943)	-65.8 (533)	DV
Explanatory variables				
Visits	Number of visits by the extension agent during the last cropping season	2.98 (3.1)	1.27 (2.8)	+
Access2	The average time required to reach (on foot) the extension agent, dry weather road, and local market (Minutes)	49.47 (28.7)	51.1 (27.9)	-
Package (dummy)	1 if the household takes part in the package and 0 otherwise	0.62 (0.5)	0.26 (0.4)	+
Impexep	Experience in growing IVs (years)	3.5 (2.1)	1.27 (1.5)	+
Educate1	Education level of the household head	1.27 (2.2)	1.48 (1.9)	+
Chat1	0 – no cht'at at all; 1 – only for own consumption; 2 – also for village sales; 3 – also for sales in the cities	1.33 (1.2)	1.26 (1.1)	+
Allfert	Fertilizer used per hectare (kg)	161.68 (515.2)	86.4 (178.8)	+
Rainfall	Rainfall distribution (3 – bad, 2- medium, 1-good)	2.21 (0.7)	2.30 (0.6)	+
Oxen	Number of oxen owned by the household	0.86 (0.9)	0.69 (0.8)	+
Inputindex	Input price index	0.99 (0.2)	1.01 (0.3)	-/+
Allquality 1	Plot quality (3 – bad, 2- medium, 1-good)	1.22 (0.8)	0.86 (0.7)	-/+
Sorgindx	Sorghum price index	1.03 (0.4)	0.97 (0.3)	+

Notes: DV – Dependent variable

[18]For users of both varieties, the GMs are the ones earned from IVs.

The variables that mainly distinguish UIVs and UFVs are number of visits by the extension agent, package participation, experience in growing IVs, fertilizer use, plot quality, and grass margin per hectare.

Table 7-2. Descriptive statistics of the variables used in the opportunity costs regression: wheat

Variable	Description	Mean (SD) - UIVs	Mean (SD) - UFVs	Expected effect on OCs
Gmboth	GM per hectare (Birr)	1731.1 (1280.4)	1015.7 (1119.3)	DV
Explanatory variables				
Age	Age of the household head (Years)	44.84 (11.9)	51.21 (15.6)	-
Scoldumy	Education level of the household head	2.14 (1.4)	1.78 (1.2)	+
Exteexpr	Experience in the extension package (years)	1.5 (1.9)	0.30 (1.1)	+
Rfdistri	Rainfall distribution (1 – bad, 2- medium, 3-good)	2.2 (0.6)	1.85 (0.6)	+
Plotqulx	Plot quality (1 – good, 2- medium, 3-bad)	2.42 (0.7)	2.51 (0.6)	-/+
Chemical (dummy)	1 if chemical is used and 0 otherwise	0.48 (0.5)	0.42 (0.5)	+
Fertph	Fertilizer used per hectare (quintals)	1.41 (0.8)	0.83 (0.7)	+
Oxenph	Oxen ownership per hectare	4.76 (5.2)	5.11 (5.03)	+
Whetinde	Wheat price index	1.01 (0.2)	0.99 (0.3)	+
Sold (dummy)	1 if wheat is sold and 0 otherwise	0.58 (0.5)	0.39 (0.5)	+
Laborph	Labor used per hectare (mandays)	192.88 (249.6)	81.07 (77.2)	-/+
Inpuindx	Input price index	0.97 (0.1)	1.02 (0.2)	-/+

Notes: DV – Dependent variable **Source:** 1999 AAU / USAID data.

7.6. Estimation results and discussions

Table 7-3 below reports the average opportunity costs generated from different homogeneous treatment statistical procedures discussed in Sub-section 7.4.1.

Table 7-3. Average opportunity cost estimates: sorghum and wheat

Method	Opportunity costs using GM in Birr /Ha. (sorghum)	Opportunity costs using GM in Birr /Ha. (Wheat)
Simple OLS[a]	168.2	435.4
Mean difference[b]	151	299
Matching	433.8	276.5
Treatment regression[d]	107.8	341.3

Source: Own results based on 2001/2002 original survey data for sorghum and the 1999 AAU / USAID household survey for wheat

[a]This is considering variety use as an exogenous variable.

[b]Users of both varieties side by side on similar quality plots are considered for sorghum. For wheat, different group of farmers (one group using IVs and the other group using using FVs) are considered. For wheat, all farmers are planting both set of varieties on plots of the same quality.

[c] The regression results are reported in Appendix IV-C (Tables IV-10 and IV-11).

Depending on the method of estimation used, average opportunity cost for sorghum ranges from 108 to 434 Birr /Ha. The average opportunity cost for wheat ranges from 277 to 435 Birr /Ha. Based on theoretical consistency, either the matching or treatment regression results can be considered to be more reliable.

If their plots have to be used for on-farm conservation purpose, the amount of compensation that sorghum farmers can expect is 272 Birr /Ha (the average GM per hectare from the IVs of sorghum). If their plots have to be used for on-farm conservation purpose, the amount of money that wheat growing farmers can expect is about 1 374 Birr /Ha (the average GM per hectare from the IVs of wheat). The compensation is higher for wheat because it is a crop that attracts better price per quintal.

Sorghum growing farmers have reported the yield from improved and local varieties of sorghum during bad and good seasons provided that they use the same levels of inputs as it is during the survey year. Computing the local and improved variety GMs during good and bad seasons reveals that UIVs gain 203.5 Birr /Ha more during a good season while they lose 121 Birr /Ha more if the season is bad. Thus during bad season scenario, UIVs are losing even more than UFVs.

All these estimates are assuming that GM from the IVs is the first best alternative use of the land allocated for the respective crops. On top of this, if IVs are subsidized, the level of subsidy should be added as a hidden cost for the purpose of national accounting calculations. In Chapter 8, all these results will be used to estimate on-farm conservation costs at the national level based on some assumptions to be spelled-out there.

In general, the mean GM/Ha for UIVs is greater than the mean GM/Ha for UFVs. However, having a closer look at the opportunity cost figures reveals that all UIVs are not equally enjoying the benefits of IVs and neither are all UFVs facing equal opportunity costs. There are UFVs who have earned a GM greater than the average of the UIVs as there are UIVs losing. If a farmer is not using IVs, it should not always involve opportunity costs because that farmer could have earned even less GM per hectare had he used IVs. By the same token, higher GM from the local varieties does not necessarily mean that there is no opportunity cost involved because even higher GM per hectare could have been earned had the farmer used IVs. Neither does loss in GM from the FVs mean that higher opportunity costs are sacrificed because even more losses could be faced from using IVs. For instance, according to the wheat data analyzed, from 50 UIVs, 12 have a net loss and from 352 UFVs, 131 have attained a GM per hectare greater than the average of the UIVs. What factors could explain such differences? Tables 7-4 and 7-5 below report FIML estimates of a switching regression model meant to address this question for sorghum and wheat, respectively.

Table 7-4. FIML regression results of a switching regression model: sorghum

Variable	Coefficient	Variable	Coefficient
Regime 1 – UIVs		**Regime 0 – UFVs**	
Constant	696.7 (1.7)	Constant	1412.6 (3.0)
ACCESS2	-6.4** (-2.3)	ACCESS2	-1.1 (-0.34)
CHAT1	72.9 (1.1)	CHAT1	94.3 (1.4)
RAINFALL	-189.6* (-1.8)	RAINFALL	-266.6*** (-2.7)
EDUCATE1	9.4 (0.25)	EDUCATE1	18.1 (0.51)
ALLQUALI	-91.1 (-0.88)	ALLQUALI	-351.3*** (-3.4)
INPUTIND	-919.5*** (-3.0)	INPUTIND	-684.3* (-1.9)
ALLFERT	0.24 (0.93)	ALLFERT	-0.23 (-0.74)
VISITS	51.0** (2.3)		
IMPEXEP	266.5*** (7.3)		
PACKAGE	342.2** (2.1)		
OXEN	79.4 (1.0)	OXEN	195.3** (2.3)
SORGINDX	598.4*** (3.2)	SORGINDX	644.8*** (3.1)
Sigma(1)	603.3 (13.0)	Sigma(0)	678.7 (10.9)
Dependent variable	GMM	Number of observations is 175	
Log likelihood function	-1440.82	Sample separation variable is USER	

Notes: ***-Significant at 1%; **- Significant at 5% and *- Significant at 10%. Figures in parentheses are the ratio of the coefficient to the estimated asymptotic standard error.

According to the results reported above, access to markets and extension, farmers' involvement in the extension package, number of visits by the extension agent, experience in using IVs, rainfall suitability, and output price index are the most important factors positively affecting the GM earned by UIVs. These results are all in line with prior expectations. For UFVs, oxen ownership, rainfall suitability, land quality, output price index, and cash crop farming have positive and significant effect on the GM earned. On the contrary, input price index is negatively and significantly affecting GMs of both UIVs and UFVs.

The GM difference between the UIVs and UFVs in the two regimes is our definition of opportunity cost. Hence, the factors that have a net positive (negative) effect on the difference between GM in regime 1 and 0 are affecting opportunity cost positively (negatively). Accordingly, access to market and extension, number of visits by the extension agent, participation in the extension package, fertilizer use, and experience in growing IVs are the most important factors affecting opportunity costs positively. On the contrary, opportunity costs are negatively related with rainfall distribution, land quality, education level of the household head, input price index, oxen ownership, sorghum price index, and cash crop farming.

The results imply that IVs have comparative advantage when accompanied by better access to markets, extension, fertilizer supply, and credit. On the contrary, better quality plots and better rainfall distribution improve the comparative advantage of using FVs.

The effect of plot quality on opportunity costs is negative implying that better quality plots have improved the comparative advantage of using FVs. In other words, unlike common expectations and presumptions of the adoption literature (Feder *et.al*, 1985; Feder and Umali, 1993), the FVs are even more advantageous with better quality plots and better rainfall than IVs. In both the sorghum and wheat data-sets, good quality plots are not skewed to improved variety (ies) use. For instance, for sorghum farmers who are using both varieties side by side, the average quality of the plots used for the improved variety is 1.24 while it is 1.25 for the plots used for the FVs. The descriptive statistics presented in Sub-section 3.6.2 also shows that farmers do not necessarily allocate good quality plots to IVs.

Cash crop farming reduces the opportunity cost because these farmers are typically better-off farmers to whom sorghum has a very little contribution to household survival. Education level of the household head reduces the opportunity cost implying that better educated farmers do not necessarily have comparative advantage in using IVs. Oxen ownership also reduces the GM difference. Thus, better-educated farmers can best use their own varieties.

The negative impact of input prices on GM is more pronounced for UIVs reflecting the capital intensity of growing these varieties. The UFVs have benefited better from better output prices. Given the better yield potential of the IVs, the UIVs have been using more inputs, incurring more costs, producing more, and getting lesser GM per hectare due to price collapse.

Table 7-5. FIML regression results of a switching regression model: wheat

Variable	Coefficient	Variable	Coefficient
Regime 1 – UIVs		**Regime 0 – UFVs**	
Constant	-770.7 (-1.3)	Constant	-699.6 (-0.42)
Laborph	1.3*** (2.6)	Laborph	-1.2 (-0.98)
Age	1.4 (0.3)	Age	30.6** (2.20)
Sold	1048.7*** (7.6)	Sold	297.9 (0.75)
Exteexpr	112.0*** (2.8)		
Inpuindx	-847.8** (-2.2)	Inpuindx	588.3 (0.46)
Scoldumy	248.7*** (4.7)	Scoldumy	260.0** (2.00)
Plotqulx	-110.5 (-1.1)	Plotqulx	398.2** (1.98)
Rfdistri	194.5** (1.9)	Rfdistri	-368.1 (-1.56)
Oxenph	.983 (0.08)	Oxenph	41.9 (1.11)
Whetinde	1727.6*** (8.9)	Whetinde	1341.9* (1.66)
Chemical	-124.9 (-0.9)	Chemical	368.6 (0.97)
Fertph	-71.8 (-1.00)	Fertph	-518.2*** (-2.6)
Sigma(1)	971.9 (26.9)	Sigma(0)	1272.4 (7.8)
Dependent variable is GMBOTH		Number of observations is 402	
Log likelihood function -3432.61		Sample separation variable is Improvex	

Source: Own results based on the 1999 rural household survey by Economics Department, AAU, in collaboration with USAID

Notes: ***-Significant at 1%; **- Significant at 5% and *- Significant at 10%. Figures in parentheses are the ratio of the coefficient to the estimated asymptotic standard error.

According to the results reported above, quantity of wheat sold, farmers' experience with the agricultural extension service, rainfall suitability, education level of the household head, intensity of labor use on the farm, and output price index are the most important factors positively affecting the GM earned by UIVs. On the contrary, input price index is negatively and significantly affecting GM from IVs. For UFVs, age of the household head, education level of the household head, land quality, and output price index have positive and significant effect on the GM earned. Fertilizer use per hectare is negatively related with GM earned from FVs.

Following the interpretation strategy used in Table 7-4, labor use per hectare, fertilizer use per hectare, farmers' experience with the extension package, farmers' participation in selling wheat, rainfall suitability, and wheat price index are the most important factors affecting opportunity costs positively. Unlike sorghum, RF suitability increases the comparative advantage of growing IVs of wheat. Age of the household head, education level of the household head, land quality, and input price index are affecting the opportunity cost negatively.

Fertilizer and labor use per hectare increase the opportunity cost of not using improved wheat varieties implying that farmers who can apply these inputs easily have comparative advantage to use IVs. IVs that are usually distributed to farmers along with fertilizer and other inputs are more responsive to fertilizer. Farmers' decision to use fertilizer more intensively with improved seeds is, therefore, rational. Farmers with better experience in using IVs and those who are selling the majority of their produce get better benefit from IVs. When output prices are more favorable, the IVs are found to have better comparative advantage.

Education and age of the household head reduce opportunity costs enabling the UFVs to earn better GM. As it is the case for sorghum, the effect of plot quality on the opportunity cost for wheat is also negative. Unlike frequent claims, the FVs are found to be more advantageous with better quality plots than IVs implying that the FVs can even have better comparative advantage with good quality plots. Oxen ownership also reduces the GM difference in growing improved and local varieties implying that farmers' capacity to undertake agricultural practices on time enables them to reduce the opportunity cost.

By way of comparing wheat and sorghum results, the negative impact of input prices and the positive impact of output prices were found to be more important for wheat. Unlike sorghum, better output prices increase the opportunity cost because better prices benefit more the improved wheat variety users as the wealthy farmers, who are able to store and sell when prices are better, are the ones predominantly using the IVs. Given that wheat is relatively a rich man's crop, the prospect for getting a better price is very high if farmers can postpone selling their products right after harvest. Unlike fertilizer, chemicals reduce opportunity costs implying that UFVs have benefited more from other non-fertilizer chemicals. This could potentially be because FVs might have been susceptible to pests, weed, and insects as a result of which users of non-fertilizer chemicals have improved their yield potential.

7.7. Summary and concluding remarks

Once farmers and localities are selected from each stratum for on-farm conservation, the step further is to determine the compensation and craft the incentive. To this end, policy needs to ask, 'What do farmers expect when on-farm conservation is in place?' This chapter has argued that their expectations and thereby the required compensations are questions of opportunity costs they are facing.

Motivated by the absence of any empirical study on the level of compensation required to induce farmers to maintain selected local varieties, this chapter has computed and analyzed the factors affecting opportunity costs. It is hoped that the results will inform policy regarding incentive design for on-farm conservation.

The factors influencing opportunity costs are meant to inform policy on the conditions calling for lower or higher compensation. Despite the existence of positive financial opportunity costs for UFVs, farmers do not automatically switch to IVs because they are attracted by other non-monetary benefits of their varieties.

The homogeneous treatment effects model results have shown that the average opportunity cost of growing local varieties for sorghum ranges from about 108 to 434 Birr (≈ € 11 to € 43)[19] per hectare while the average opportunity cost for wheat ranges from about 277 to 435 Birr (≈ € 28 to € 44) per hectare. While the opportunity cost of land for sorghum growing farmers is about 803.76 Birr (≈ € 80) per hectare, the opportunity cost of land for wheat growing farmers has been about 1374 Birr (≈ € 137) per hectare.

Even though the over-all net financial benefit of use of the IVs is better than the FVs, non-use of IVs does not always involve opportunity costs. In other words, UIVs are not equally enjoying the benefits of IVs. By the same token, all UFVs are not facing equal opportunity costs. There are UFVs who have earned a GM greater than the average of the UIVs as there are UIVs who have earned a GM greater than the average of the UFVs. Higher GM from the FVs does not necessarily guarantee absence of opportunity cost because even higher financial benefits could be earned from the IVs. Neither does loss in GM from the FVs necessarily mean that higher opportunity costs are sacrificed because even higher losses could be incurred with IVs.

The size of opportunity costs also differs from one crop to another. The results are relevant not only for on-farm conservation policy but also for targeting improved technology adoption because they show the comparative advantages of the respective varieties. The computed opportunity costs will

[19] In July 2003, € 1 ≈ 10 Birr.

further be taken up in Chapter 8 to estimate national level on-farm conservation costs. The details of the conclusions and policy implications of the empirical results of this chapter are presented in Chapter 9.

8. Thinking beyond the boundary of farmers: national level on-farm conservation costs and the institutional incentive design in Ethiopia

8.1. From household level empirical results to issues of national interest

It is not still clear what institutional arrangement can effectively promote conservation and sustainable use of crop diversity (Bhattarai and Hammig, 1998). The existing institutional framework (policies, property rights, rules and norms) fails to force those individuals or agencies to pay the full costs of their exploitation and to reward those who are contributing to its production. This can be cited as a major reason for biodiversity loss. Given the current institutional set-up, it is difficult or intractable to charge those responsible for the loss and reward those responsible for its maintenance. Crafting enforceable institutions for on-farm conservation is, therefore, a vital step to implement sustainable conservation strategies. This chapter, heavily drawing from the results of previous empirical chapters, provides a framework for the required institutional innovations for better crop diversity and development outcome.

Both household related contextual factors and national endowments are equally relevant to undertake on-farm conservation of crop diversity. *In-built incentives*, preferences, and opportunity costs are important at the household level. Moreover, policy has to address issues of national interest: conservation costs, institutional gaps, and harmonizing conservation with the rural development pathway to be followed. Establishing an institutional framework at the local, national, and international levels is indispensable for raising finance, reducing the trade-offs involved, harmonizing conflicts of interest, and monitoring the outcome. In addition to acknowledging farmers' contribution in maintaining crop diversity, institutional development is needed to establish an efficient exchange system (Virchow, 1999a).

The conservation of crop diversity for crop improvement has emerged as a controversial policy issue in recent years (Gollin *et.al*, 1998). Much of the controversy surrounds the trade-offs involved among the elements of the critical triangle: economic growth, poverty alleviation and environmental sustainability. As it has been argued in previous chapters, the empirical results of the thesis can serve an important role in the institutional design for on-farm conservation of crop diversity. This, among other things, requires considering the objectives and endowments of the respective stake-holders involved in on-farm conservation.

Before designing policy incentives, policy makers have to address different questions. Having this in mind, this chapter attempts to contribute to the following questions:

- How much does on-farm conservation cost and what factors determine the magnitude of the cost?
- How can policy incentives be designed so that they meet conservation goals and farmers' needs (goals, concerns and constraints)?
- How can the government link conservation of crop diversity with variety utilization and agricultural development?
- How can policy use farmers' organizations or local informal institutions for on-farm conservation purpose?
- What is the operating cost of gene bank conservation?

While blending these issues and questions, there are three-fold purposes in this chapter: estimate on-farm conservation costs at the national level using micro-level opportunity cost data and present the annual operating costs of ex situ; describe the stake-holders in on-farm conservation incentive design; and lastly present the possible institutional incentive scheme to undertake on-farm conservation of crop diversity in Ethiopia using coffee, sorghum and wheat as examples. All of these objectives are addressed building upon the empirical results of Chapters 3 and 5 through 7. Of particular interest are the opportunity cost computations of Chapter 7.

To address these questions, the chapter begins with an account of the factors that need to be considered when undertaking a decision on the optimum level of investment for on-farm conservation. This part is meant to elaborate on the variables that should be taken into account to decide on the optimum level of investment for on-farm conservation. Section 8.3 takes up the estimation of on-farm conservation costs based on the empirical results of Chapter 7 and presents estimates of ex situ operating costs. Sections 8.4 and 8.5 are devoted to the issues of institutional incentive design building upon Chapters 5 through 7. Lastly, Section 8.6 concludes presenting different policy incentives based on farmers' in-built incentives and opportunity costs.

8.2. Factors influencing the optimum level of on-farm conservation investment

How is the 'optimum' level of investment for on-farm conservation determined? The word 'optimum' is put within quotation marks because it refers here not to the conventional optimum based on equality of marginal cost and marginal benefit but to the level of on-farm conservation investment that has to be decided upon considering various factors beyond cost and benefit. Without substituting one conservation strategy for the other (see Section 2.5), policy makers have to decide on the 'optimum' level of in situ and ex situ conservation and the amount of direct and opportunity costs required to arrive at the 'optimum' level of conservation.

Neither on-farm (except *de facto*) nor ex situ conservation strategies are possible unless the country is prepared to incur costs. Since the costs to be incurred on ex situ and on-farm conservation are of different nature, there are different factors influencing the cost requirements. Ex situ costs are basically dependent on the size and number of collections to be made, the size of the establishment, the cost of capital establishment, the collection cost involved, the variability of the facility requirement of the different accessions, and the rate of replacement required for the different accessions. If the government has the required policies to deal with the potential health and environmental risks, the emerging biotechnology is said to offer tools that can be used to minimize the cost related with ex situ conservation in the cold room gene bank (Yibrah, 1998).

When it comes to on-farm conservation, in addition to farmers' opportunity costs that have to be compensated, there are administrative costs including negotiation, supervision, and establishment costs. These costs could be high especially during the first few years until the system is adapted to the way of life of the local community. The 'optimum' level of on-farm conservation and its costs are dependent on the number of agro-ecologically distinct hot-spots identified or the diversity of the farming system, the amount of ex situ conservation under way, the size of land required, the number of varieties (accessions) to be maintained, the level of *de facto* conservation under way, the extent of hybrid variety 'replacement' under way, the number of replications required, and other negotiation and preparatory costs. The other factor that needs to be considered is the suitability of technological innovations and the rate of technological change that occurs through breeding and biotechnology. The more suitable the technologies are to farmers and the faster their turn-over, the more costly it is to maintain indigenous varieties on-farm for the sake of crop diversity because this trend increases the foregone opportunity cost.

The size of land required and its opportunity costs are the other important factors determining the on-farm conservation cost. As the land size required gets larger, the cost will be scaled up accordingly even though there could be some *economies of scale* potentials. The opportunity cost is a function of the productive capacity of the land in its first best use. The larger the number of agro-ecologically representative hot-spots identified by policy makers in consultation with scientific assessment, the higher the number of on-farm conservation sites that have to be established and then the higher will be the national level on-farm conservation costs.

The nature of the crop (cross pollinated versus open pollinated[1], high value cash crop versus low value food crop, and/or perennial versus annual crop) is another factor that determines the on-farm conservation cost due to differences in requirements, opportunity costs, and number and complexity of contracts.

If there is any communality between the outputs of ex situ and on-farm conservation, higher level of ex situ conservation will lower the costs of on-farm conservation that have to be incurred as there is no need to duplicate efforts. On-farm conservation costs are also affected by the extent of replacement of FVs by IVs and the nature of IVs being distributed which could be hybrids or just simple selections. If what is occurring is replacement than addition, the required on-farm conservation investment will be high. If the IVs substituting FVs are hybrids, the required on-farm conservation investment will, accordingly, be high.

The level of *de facto* conservation under way by farmers is the other issue to be considered while making decisions on the level of on-farm conservation investment. As shown in Chapters 5 and 6, there are a lot of FVs conserved year after year by farmers without any policy inducement. Thus, *de facto* conservation can save considerable amount of conservation cost that should otherwise be incurred.

There are three types of cost concepts relevant for on-farm conservation: *transaction costs*, *opportunity (economic) costs*, and *financial costs*[2]. All the three types of costs should be considered in crafting institutions to arrive at

[1]If a crop is open pollinated, the rate of change in its diversity on-farm will be high.

[2]The three cost concepts are not mutually exclusive.

the efficient way of conservation of crop diversity. The transaction costs and difficulty of incentive design are, in turn, functions of the diversity of the agro-ecology, absence or presence of established informal institutions and farmer organizations, the diversity of farmers' expectations, and the institutional solution adopted. In this regard, one of the challenges of farmers' heterogeneity for on-farm conservation is that it makes the transaction cost of negotiation and incentive design more difficult.

The costs that are computed below (Section 8.3) are considering only the *ex ante* compensation costs based on the computed opportunity cost at the farm household level. For a more comprehensive assessment and comparison with ex situ costs, there is a need to include other administrative and regulatory costs based on the on-farm conservation schemes in place. To this end, monitoring the costs of the GEF-sponsored on-farm conservation project can offer live experience.

8.3. On-farm conservation cost and ex situ operating cost estimates

Costs of crop diversity conservation are becoming of considerable interest in the crop diversity literature (Burstin *et.al*, 1997; Epperson *et.al*, 1997; Virchow, 1999a; 1999b; Pardey *et.al,* 1999). This interest is partly motivated by the need to address issues of financial requirements, designing efficient conservation strategies, and analyzing factors determining the financial requirement. However, almost all these studies are at aggregate level and primarily for ex situ conservation or just for all types of conservation activities without making any distinction. The more interesting theme could be analyzing on-farm conservation costs and the factors determining the efficiency of the respective conservation strategies so that policy can optimize whenever it is feasible. Of course, to compare the cost efficiency of different conservation strategies, one has to make sure that the crop diversity outcome from different conservation strategies is the same.

Estimating costs of on-farm conservation at the national level is done aggregating the opportunity costs computed in chapter 7 and taking the land requirement information from plant breeders during the survey. According to the information gathered from breeders, a hectare of 3 replications is necessary to maintain a single crop landrace. Experience from the Ethiopian pilot on-farm conservation projects also shows that optimum farm size for maintaining a population should be similar to the farm size the farmers usually allocate to the different crops. For this reason, a hectare of land with three replications is taken as a reference to compute national level on-farm conservation costs.

8.3.1. National level on-farm conservation cost estimates using farm level opportunity costs

8.3.1.1. On-farm conservation costs for sorghum

To estimate national level on-farm conservation costs, the average opportunity cost results (Table 7-3) are taken. If 100 landraces of a crop have to be maintained at the national level, requiring three replications of 1 hectare each, annual opportunity cost compensation ranges from 32,400 Birr (€ 3,240) to 130,200 Birr (€ 13,020) per annum[3]. Based on these estimates, maintaining one landrace of sorghum would cost the government between 324 Birr (€ 32) and 1,302 Birr (€ 130) annually. Using the mean difference and the simple OLS results, the in-between annual compensation costs will range from 45,300 Birr (€ 4,530) to 50,400 Birr (€ 5,040). The cost of conserving a landrace on-farm will, then, range from 453 Birr (€ 45) to 504 Birr (€ 50), respectively.

These cost estimates are with the understanding that farmers harvest the yield from the varieties maintained on-farm and apply all the inputs and undertake cultural practices by themselves. As the contract is supposed to be concluded in advance, no farmer is expected to exercise *moral hazard* behavior to productively use the varieties targeted for on-farm conservation. If the contract is based on the yield outcome, there will be moral hazard problem as it has been observed from the GEF-sponsored on-farm conservation project in Ethiopia (Demissie, and Arega, 2000).

What if policy makers would like to rent farmers' land for the same purpose? To this end, GM per hectare from the IVs has been computed considering all costs except land. This computation has resulted in the GM per hectare of 272.46 Birr (€ 27) that farmers can expect to get as a compensation for their land. This is rather low (even lower than the higherst opportunity cost estimate) but it has to be noted that all farmers' labor costs (paid out and imputed) have been included and the GM is a function of the

---

[3]To know the exact national cost of on-farm conservation, we have to know the number of on-farm conservation sites that have to be established. That needs assessing the number of landraces requiring conservation. We are taking 100 landraces for the sake of illustration. The unit cost per landrace is, however, not affected by this choice.

level of agricultural productivity and output prices which are low at the time of the survey. Using 300 hectares of farmers' land for on-farm conservation purpose, accordingly, requires a compensation cost of 81,600 Birr (€ 8,160) per annum. In other words, maintaining one landrace of sorghum would cost 816 Birr (€ 82) annually. To be compared with the opportunity cost results above, the GM to be earned from sorghum variety planted (for policy purpose) has to be computed. Moreover, the costs of establishing and running the on-farm conservation schemes (compensating opportunity costs and renting farmers' plots) should be computed and compared in the two cases.

The other alternative to compute on-farm conservation costs is to use the market value of land. In the absence of official land market for policy reason, sampled farmers were interviewed on their willingness to accept if they rent out a hectare of land and their willingness to pay if they rent in a hectare of land. The results of this exercise are reported in Appendix 3 (Table III-2) [4].

The descriptive results show that there is a huge range of difference across villages. Land quality (land productivity) differences across villages, population density, farmers' heterogeneity in resource endowment, farmers' heterogeneity in land holding, and the opportunity to use the land for high value crops could be the reasons. For obvious reasons, farmers' willingness to accept for a hectare of land (used in estimating on-farm conservation costs) is consistently higher than their willingness to pay. The lowest value of farmers' willingness to accept for a hectare of land is 343 Birr (€ 34) and the highest is 707 Birr (€ 71). Payments for farmers plots should, therefore, range from 343 Birr per hectare per annum to 707 Birr per hectare per annum. For 100 landraces planted on a hectare with three replications, the annual cost of compensation for land ranges from 102,900 Birr (€ 10,290) to 212,100 Birr (€ 21,210) per annum. In other words, the average compensation costs per landrace will range from € 103 to € 212 which is about one-third of a payment made to a hectare of land.

---

[4] In drought prone areas, land is relatively cheap to the extent that most of these farmers were found to be willing to give a certain portion of their land without necessarily requiring any payment.

8.3.1.2. On-farm conservation costs for wheat

Maintaining the assumptions as in the case of sorghum, annual compensation cost for wheat ranges from 82,950 Birr (€ 8,295) to 130,620 Birr (€ 13,062). These estimates imply that the cost of maintaining one landrace of wheat would range from 830 Birr (€ 83) to 1,306 Birr (€ 130) annually. The upper limit figures are almost the same for sorghum and wheat. The lower figures imply that, compared to sorghum, the on-farm compensation cost is higher for wheat by about 156 percent. Adopting the mean difference and treatment regression results, the in-between annual compensation costs will be 89,700 Birr (€ 8,970) and 102,396 Birr (€ 10,240), respectively. In this case, the unit cost of a landrace will, respectively, range from € 90 to € 102.

GM per hectare from IVs of wheat is used as the first best use of land allocated for growing wheat. If their plots have to be used for on-farm conservation purpose, the compensation for wheat growing farmers is about 1,374 Birr (€ 137) per hectare per year. This is the cost of renting wheat farmers' land. This implies that the annual on-farm compensation cost is 412,200 Birr (€ 41,220). In other words, maintaining one landrace of wheat would cost 4,122 Birr (€ 412) annually. This is about 5 times higher compared to the case of sorghum. The compensation is higher for wheat because it is a crop that commands better price and hence the opportunity cost of land that would be allocated for wheat is higher.

8.3.1.3. On-farm conservation costs for coffee based on farmers' valuation of land

The on-farm conservation cost for coffee is estimated based on farmers' valuation of land. The lowest value of coffee farmers' willingness to accept for a hectare of land is 346 Birr (€ 35) and the highest is 803 Birr (€ 80). These figures do not show the real worth of land in coffee growing areas. Had it not been for the collapse of the price of coffee, the difference between value of coffee land and land allocated for other cereals like sorghum would have been more than what the figures suggest. Making the same assumptions as in the case of sorghum, the annual cost of on-farm conservation ranges from 103,800 (€ 10,380) to 240,900 Birr (€ 24,090) per annum. The annual cost per unit of landrace ranges from 1,038 Birr (€ 104) to 2,409 Birr (€ 241). This is only about 14 percent higher than the case of sorghum.

8.3.1.4. Points of cross-cutting importance

Actually, there could be a lot of potential to have *economies of scale* effects by maintaining more than one local variety on a single plot. This very much depends on the varieties identified in the conservation site, their differences in resource requirement (weather, soil type, fertilizer, and labor), their similarity in growth pattern, and their prevalence in the same locality. In all the computations of this chapter, it is assumed that each landrace needs separate plots. Hence, the estimates could be considered as upper bounds from this perspective.

It is well known that IVs are mostly subsidized. Subsidies artificially make IVs comparatively more attractive compared to FVs. By increasing the opportunity costs, subsidies are artificially calling for higher compensation for on-farm conservation. Reducing the subsidies or avoiding them will relieve the government's budget and enable the FVs and IVs compete freely (Prakash and Virchow, 2003). Thus, the computed on-farm compensation costs based on opportunity cost are marginally inflated to the extent that IVs are subsidized. However, as the subsidies are flat rates, our analysis of the factors affecting the opportunity costs in Chapter 7 will remain to be valid.

Are all these costs affordable to Ethiopia? To answer this question, the relative importance of the respective crops and the public goods nature of the resources maintained (not only to Ethiopia but also to other countries) will have to be considered. To give insight on the importance to the Ethiopian agriculture of the crops considered, table 8-1 below reports the value of annual estimated production of sorghum and wheat, and the value of annual coffee exports.

Table 8-1. Gross value of coffee sent to central auction and gross values of sorghum and wheat produced in Ethiopia

Year	Gross value in Birr (in 000's)		
	Coffee*	Sorghum**	Wheat**
1981/82	282,534	725,737	318,651
1982/83	292,846	614,880	389,604
1983/84	264,606	1,409,155	558,427
1984/85	185,174	620,248	585,121
1985/86	340,636	644,849	427,130
1986/87	247,863	475,888	335,093
1987/88	250,728	477,057	402,229
1988/89	317,087	414,657	413,317
1989/90	262,621	543,705	399,446
1990/91	258,829	744,359	647,562
1991/92	216,358	625,947	623,130
1992/93	619,682	612,192	761,026
1993/94	1,422,484	675,544	651,782
1994/95	1,679,798	1,336,761	1,038,107
1995/96	1,520,723	2,074,866	1,336,576

Source: **Computed based on yield, cultivated area and price data from CSA bulletins. *Computed based on data from Coffee Statistics Handbook (Ministry of Coffee and Tea Development, 1984) and CTA various issues.

It can be concluded that given their public goods nature[5] and their additional insurance and breeding value, investing on on-farm conservation of the diversity of these crops pays off. Comparing the figures in Table 8-1 with the on-farm conservation costs suggests that the costs are small proportions (less than a percent of the gross value) of the monetary values of the respective crops. The compensation costs for both crops are low compared with the importance of these crops with annual production of over 12 million quintals per annum each (NSIA, 2001). On top of this, the country can get multiplier and scale effect benefits in other sectors if it manages to improve agricultural production and productivity using its genetic resource stock.

EARO (the then Institute of Agricultural Research) has been releasing IVs of sorghum since 1970. Provided that the right cultural practices are applied, the IVs of sorghum have a yield potential of 14Qt/ha to 50Qt/ha on farmers' fields with an average of 23.5 Qt/ha (NSIA, various issues). This is in contrast to the national average of 12 Qt/ha (CSA statistical bulletins,

[5]This makes extra unit cost of producing genetic information as small as the economy can utilize them in a large scale.

various issues). Since 1966, breeders have released many IVs of wheat with a yield potential on farmers' fields ranging from 15Qt/ha to 42.5Qt/ha with an average of 31 Qt/ha. This contrasts with an average national yield of 10 Qt/Ha. All these agricultural productivity improvements are derived from research undertaken in agriculture using FVs as the major inputs in plant breeding. Consequently, not only Ethiopia but also other benefiting countries and organizations have to invest in the conservation and sustainable use of these resources.

8.3.2. Ex situ operating costs

In cold room gene banks, fixed costs constitute a high proportion of total costs. Benefits of *scale economies* are likely to prevail through cost complementarities resulting from shared facilities for maintaining different types of germplasms. For instance, consolidating maize and sorghum collection in a shared facility may offer the prospects of significant cost savings compared to maintaining each collection in a separate facility.

Secondary data were collected from the IBCR on operating costs[6] and holding capacity of the cold rooms. The average cost of maintaining a landrace in the cold room was computed using these data and data taken from previous studies on the number of accessions held. Table 8-2 below reports the results of this exercise.

---

[6]The cost items included are hired labor, chemicals, laboratory equipments, vehicles, storage, perdiem, transport, fuel, maintenance, electricity, water, and daily labor. Laboratory equipment and vehicle costs are apportioned to the respective years based on their estimated service life and using straight line depreciation method.

Table 8-2. The holding capacity and operating costs of the Ethiopian gene bank

Year	Operate	Capacit	Hold	Percent	Fulcost	Coshold	Reduce
1990	1,424,297	103,860	37,169	35.79	13.71	38.32	179.4
1991	835,067	103,860	37,169	35.79	8.04	22.47	179.4
1992	763,587	103,860	37,169	35.79	7.35	20.54	179.4
1993	943,312	103,860	37,169	35.79	9.08	25.38	179.4
1994	1,053,426	103,860	54,000	51.99	10.14	19.50	92.3
1995	842,804	103,860	54,000	51.99	8.11	15.61	92.3
1996	1,190,694	103,860	54,000	51.99	11.46	22.05	92.3
1997	1,100,916	150,840	54,000	35.80	7.30	20.39	179.3
1998	2,352,147	150,840	55,548	36.83	15.59	42.34	171.6
1999	3,133,865	150,840	55,548	36.83	20.78	56.42	171.6
2000	2,638,417	150,840	55,548	36.83	17.49	47.50	171.6
2001	2,977,393	150,840	61,000	40.44	19.74	48.81	147.3

Notes: Operate, Capacit, Hold, Percent, Unitcost, Fulcost, Coshold, and Reduce refer to estimated annual operating cost (birr), total holding capacity (in number of accessions), number of accessions held, percentage of the full capacity of the gene bank currently utilized, cost per accession at full capacity (birr), cost per accession at current holding (birr), and percentage of cost reduction possible at full capacity, respectively. The crop considered to estimate the holding capacity is sorghum, single line, medium seeded.

Source: IBCR documentation; Yibrah, 1998; Worede, 1991; and Virchow, 1999a.

The second column shows the annual estimated operating costs for the respective years based on data collected from IBCR. Accordingly, the annual operating cost ranges from 0.76 million Birr to 3.13 million Birr. Percentage of full capacity utilized is computed using the number of accessions held and the holding capacity in the respective years.

The data by and large show that the cold room gene bank is still not able to operate at its full capacity. On average only 36 to 52 percent of its capacity is utilized during the years 1990 to 2001. If the country manages to utilize its gene bank at its full capacity, there is a huge potential to maintain a lot of accessions without incurring costs for additional gene bank facility. By utilizing the gene bank at its full capacity, it is possible to reduce the unit cost of maintaining an accession by about 92 to 179 percent. Currently, only 40 percent of its full capacity is utilized.

The cold room gene bank conservation cost depends on the nature of the seed which refers to its size and technology requirement to make it stay viable. While the size affects the holding capacity of the cold room gene bank, the technology requirement affects the cost of the technology. Both of these variables will affect the unit cost of maintaining an accession.

How reliable are the on-farm conservation costs estimated in Sub-section 8.3.1 and how comparable are they with the ex situ conservation costs in the preceeding table? The cost estimates are meant for illustration and they are as good as the plausibility of the assumptions made. There is a possibility to over (under) estimate the costs. It is only since 1998 that the IBCR was established as an independent Institute with mandate on both plant and animal genetic resources. The gene bank was under MoA and then under EARO. Hence, getting independent cost estimates was very hard. Moreover, some of the operating costs are shared with in situ conservation activities. Since the outcomes are different (see Section 2.5) and the on-farm conservation cost estimates are *ex ante* (unlike the *ex post* gene bank costs), the comparison in terms of cost per unit of accession is not pursued. However, both the gene bank and on-farm conservation costs are relevant in their own right for the purposes discussed at the beginning of Section 8.3. The on-farm conservation cost estimates illustrate how opportunity cost estimates can be used to estimate on-farm conservation costs. The gene bank conservation costs illustrate the operating costs of running the gene bank and the extent to which it is run at its full capacity. Despite their limitations, both cost estimates still show the range of magnitude of costs to maintain an accession ex situ and in situ.

8.4. The stake-holders in the incentive design and institutional linkages

Institutional research concerning on-farm conservation is a topic of recent interest and past neglect (Wood and Lenne, 1997). Addressing Ethiopia's critical environmental and resource degradation problems requires, among other things, better understanding of the institutional structures, gaps, and required innovations. In doing so, establishing market incentives, removal of perverse incentives, and establishing grass-root support for farmers in maintaining local crops is indispensable (Brush, 1991). At the international level, creating the linkage involves establishing partnerships among organizations supplying and demanding (utilizing) genetic resources (Mugabe and Ouko, 1994).

Local community (farmers), government, NGOs, businesses, and the international community are the institutional actors in crop diversity conservation. The purpose of participatory rural development approaches like on-farm conservation is to integrate farmers' tastes and preferences in the process of designing the intervention. Hence, to be sustainable, any conservation intervention must involve farmers as major stakeholders.

As far as incentive design is concerned, it is imperative to address the following questions. On what basis should the government set the level and type of compensation to farmers? What factors determine the decision to follow either a fixed payment schedule or flexible incentive schemes? Should the government follow long-term or short-term contracts[7]? When is it better for the government to delegate on-farm conservation activities to farmers and when is it preferable to get it done by the government or its agency?

For those 'pro-diversity' farmers who are already conserving crop diversity (like the ones identified in Chapters 5 and 6), there is no strong incentive to act against the interest of the conservation agency unlike those who are not maintaining local varieties *de facto*. The 'pro-diversity' farmers are paying lower opportunity costs (or even not at all) unlike the farmers with little in-built incentives to maintain crop diversity. Once farmers' in-built incentives are understood (Chapters 5 and 6), the other subsequent step is to compare farmers' livelihood in the 'with' and 'without' the implementation of the on-farm conservation scheme (Chapter 7) and put in place the required incentives depending on the welfare losses farmers are facing in the 'with' case.

In the design of on-farm conservation incentives, there are different stake-holders that need to coordinate and harmonize their efforts. By grouping actors into their respective activities, Virchow (1999b) has identified five major actors in crop diversity conservation which he called the '*Pentagon of crop diversity conservators*': farmers, public conservation agents at the national level, private breeding companies[8], regional and international gene banks, and local conservators. Accordingly, the incentive design should involve besides the IBCR and the farmers, local and national government organizations, private companies (pharmaceutical and biotechnology firms, and national or international private seed companies), local community groups and informal institutions, research organizations, and NGOs. Figure 8.1 below depicts the possible interaction among the potential stake-holders in the incentive design and the factors involved.

[7] *Ceteris paribus*, long-term contracts may be called for perenial crops like coffee.

[8] Public and NGO organizations and actors in the demand side (biotechnology firms and breeders) will play an important role in financing on-farm conservation.

Figure 8-1. Institutional interactions among the stake-holders in the design of on-farm conservation scheme[9]

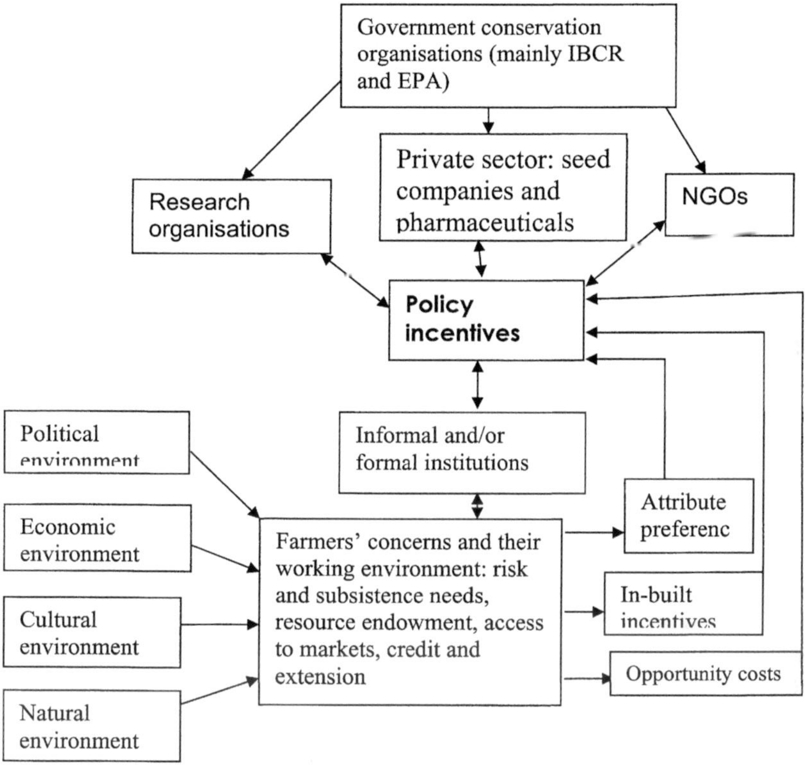

Those organizations operating above the policy incentives are the ones demanding crop diversity and/or paying the compensation. Those organizations below the policy incentives are the ones supplying crop diversity and/or demanding compensation. The flow of the chart starts with farmers' working environment in the left bottom of the scheme. The schematic presentation shows that the political, policy, economic, cultural and natural settings along with farmers' contextual working conditions

[9]When placing the organizations in the scheme, we are not considering institutions as organizations but institutions are implied by the rules governing the interactions among the organizations involved (North, 1990).

determine their in-built incentives (Section 5.2), attribute preferences (Section 6.2), and the level of opportunity costs (Section 7.2) they face.

Depending on farmers' in-built incentives, attribute preferences, and opportunity costs, policy incentives will have to be set preferably using the existing informal local institutions as intermediaries to reduce transaction costs. Government conservation organizations (mainly the IBCR and EPA) have to set their conservation goals (for instance, which crop varieties to be maintained and how much of each?) and negotiate with other stake-holders (research organizations, the private sector, and NGOs) to satisfy farmers' expectations contextually. This will result in a design of area specific and flexible '*policy incentives*'.

Interaction among the stake-holders should be coordinated to compromise trade-offs and resolve conflicts of vested interest. The level and type of payment should be set based on the opportunity costs involved. Transaction costs, legal matters involved, and equity issues will determine the flexibility of the types and magnitudes of the incentive schemes. Moreover, the responsible conservation organizations (in most of the cases the public conservators) have to monitor the levels of compliance by the respective stake-holders *i.e.* the degree to which farmers and their organizations respect and adhere to the existing constraints, both formal and informal.

Moreover, effective implementation may call for setting legal pre-requisites in place that demands creative institutional innovation. In any case, what policy expects from farmers and what farmers expect from policy should be set and negotiated contextually. The results of the empirical chapters have shown those contextual factors worth considering in the process of negotiation.

In general, there are three groups of stake-holders in the above scheme – the net suppliers of crop diversity (e.g. farmers), the net demanders of crop diversity (e.g. public and private research organizations, biotechnology companies, and pharmaceuticals) and public conservators (IBCR, EPA and NGOs). Hence, the incentive design should involve the three contracting parties. The conservators should be responsible to solicit funds not only from local net demanders but also from other national and international organizations. Due to their long lasting role in the on-farm conservation exercise, the rest of the stake-holders have to be informed on incentives and opportunity costs so that policy can face the challenging task of finding the optimal mix of '*policy incentives*'.

The empirical results of this thesis (Chapters 5 and 6) are meant to improve farmers' compliance. The information problem for the government in designing incentives is solved if policy is informed as to why or why not farmers behave in a certain way. Farmers' choices to cooperate or defect are contingent upon the extent of the satisfaction of their expectations by the government. The beauty of computing and analyzing opportunity costs is to inform the government on the 'optimum' compensation to enhance compliance by farmers.

The field research during the cropping season - 2001/2002 - reveals that farmers have organized themselves under the informal local institutions. For a successful on-farm conservation scheme, the incentive design should, therefore, involve the indigenous informal institutions. Such institutions can be used either to select farmers for on-farm conservation or to manage crop genetic resources as common-pool resources (CPRs).

8.5. Potentials of informal institutions for on-farm conservation

Both the *CBD* and *Agenda 21* acknowledge the importance of the participation and support of indigenous and local communities to the conservation and sustainable use of biological diversity. Does considering crop diversity as CPRs represent a viable policy option? This question needs an empirical investigation based on practical case studies. The intention here is to present the potential role of informal institutions that have managed to survive many difficulties in the rural areas of Ethiopia.

To make on-farm conservation more long-lasting and to reduce costs of negotiation and contract formation, the incentive structure could involve not just individual farmers but communities and the institutions governing them (See Figure 8-1). To this end, there are indigenous informal and formal institutions established by the local community to discharge different socio-economic activities (including labor sharing, rural finance, funeral, and other self-help institutional arrangements)[10]. Instead of dealing with individual farmers, these institutions can be used as mediators to operationalize the design of on-farm conservation activities for transaction cost reasons.

On top of reducing transaction costs, there could be additional reasons for considering conservation of crop diversity in a CPRs framework. Among

[10]Informal institutions play key net-working functions in the surveyed villages and almost all of the farm households are involved in at least one informal institution.

these, difficulty of establishing private property rights (see Sub-section 4.3.2, lack of knowledge of farmers about their ownership rights, *economies of scale* effects, simplicity for implementation, suitability for benefit sharing, possibility to give farmers power to materialize their rights, and spillover effects of organizing farmers for CPRs management (awareness creation, social capital, *etc.*) are the most important ones. That is why McNeely and Vorheis (2000) have considered conservation of biological diversity in a CPRs management framework.

Further more, involving the already existing informal institutions does not need creating brand new institutions but building upon the potentials of the already existing ones. It is also more sustainable as these institutions have been enduring in the face of many difficulties. Since farmers have high confidence and level of participation in these institutions, considering crop diversity as CPRs can attract many farmers' interest to participate in community-based conservation activities. Given the problems set out in Sub-section 4.3.2, instead of trying to attribute crop diversity to particular farmers, group of farmers can be given the responsibility and the policy incentive to maintain indigenous varieties. Moreover, these institutions can serve as channels of communication and facilitate interactions between the conservators and the local community. These institutions can serve intermediary role in selecting sites, securing land, negotiating with farmers, and designing incentive schemes. Then the farmers involved can store, exchange, and get access to the seeds maintained. The community gene banks described in Sub-section 2.4.2.3 are very good examples to use informal institutions for on-farm conservation purpose.

Even though using the locally operating informal institutions has these advantages, in the absence of adequate contextual experience, it is also imperative to note the following problems. One of the challenges is adapting the rules of operation of the already existing informal institutions. Due to the diversity of the types of informal local institutions in place, policy makers have to study the nature of each and identify the institutions that can best be adapted. Secondly, all farmers are not members of the informal institutions. If only members of informal organizations are considered, this may exclude some pro-diversity farmers. Last but not least, because of severe rural poverty prevailing in most parts of the country, some of the informal institutions (especially those that involve fee payment) are disintegrating. Thus, rural poverty makes using informal institutions for conservation purpose more challenging and less sustainable.

8.6. Summary, conclusions and policy implications

On-farm conservation is a process of negotiation with local farmers who are said to be custodians of crop diversity. The basic building block is that farmers should take part not as just passive seed suppliers or maintainers but as active conservationists benefiting from on-farm conservation, influencing conservation goals, and negotiating with policy makers. That is why Cromwell and van Oosterhout (2000) have pointed out that 'for conservation policies to be effective in maintaining on-farm crop diversity, knowledge of farmers as derived from resource management practices should be seen as management of shifting boundaries created by economic, socio-cultural, and environmental variables rather than as management within boundaries set by a technology package'.

To be successful, on-farm conservation has to get not only public but also local support. Envisaging such an on-farm conservation intervention, the incentive design should fulfill the different desirable features of incentive measures, namely, predictability of impact, equity, political acceptability, economic efficiency, adaptability, and administrative feasibility (Young and Cunningham, 1997; quoted in OECD, 1999). In designing the incentive measures, policy makers have to assess the respective measures with respect to each of these criteria and decide.

If the conservation objective is to motivate farmers to maintain more than they currently do, any potential regime should not limit itself to compensation for lost income (Correa, 2000). This point re-enforces the need to integrate financial incentives with other tangible non-financial benefit schemes that can convince farmers to integrate crop diversity conservation to their livelihood strategies. If farmers value the importance of local varieties to their livelihoods, they may not necessarily need policy incentive to undertake *de facto* conservation. Thus, doing nothing is one option in on-farm conservation.

In the absence of a single incentive scheme that works in all contexts, it is optimal for the government to have a menu of incentive schemes to different farm household types and localities. This chapter is concluded by outlining the required policy incentives based on in-built incentives and opportunity costs. Depending on the features of the farm household types (Chapters 5 and 6), levels of opportunity costs and the prevailing socioeconomic factors (Chapter 7), either of the following incentive schemes and/or their combinations can be put in place:-

- *Enhancing the productive potential of FVs:* This is an option that can be more plausible for farmers found in less accessible areas. These farmers are maintaining a huge stock of crop diversity. The incentive scheme works in such a way that the targeted farmers are planting the local varieties and government is assisting in improving the productive capacity of the farmers' resources (land, labor, and FVs). This could involve improving the supply of credit, fertilizer, and chemicals provided that these inputs can enhance the productivity of FVs.

- *Using farmers' land for on-farm conservation:* This scheme calls for compensating the opportunity costs of the land used for the crop under consideration. This could be the net benefit from the most profitable variety on that plot or farmers' willingness to accept for land. This is a scheme that can be negotiated (through auction, for instance) with farmers depending on the foregone net-benefit by farmers from that piece of land as derived from their willingness to pay or the net return foregone from that land. It is expected to work better in more accessible areas with better informal land markets in place. The share cropping arrangement used by farmers can also be used if policy makers are ready to deal with the contractual arrangements necessary.

- *Compensating farmers for opportunity costs*: This involves comparing GMs of what the farmer is planting with what he/she is expected to plant. This is a scheme whereby the government reimburses the opportunity cost faced by the farmer in changing the variety choice decision behavior and the farmer plants the indigenous varieties worth maintaining. The levels of compensation should depend on the levels of opportunity costs. Whether to use spatially differentiated incentive schemes or flat rate compensation (using the highest opportunity cost estimate) depends on the relative size of the transaction costs of the differentiated scheme and the extra financial cost of the flat rate. Provided that government can solicit adequate funds, this incentive scheme could be put in place in more accessible areas endangered by potential diversity loss.

- *Premium payment to farmers' varieties*: This is one step in recognition of farmers' role in safeguarding crop diversity. The initiative could be started by identifying unique varieties maintained by farmers. These are varieties that are not commonly found in other farmers' fields. This can be followed by rewarding the selected farmers. Occasionally, the IBCR and research Institutes can organize a 'seed exhibition day' during which each farmer will bring his/her own local seeds in use and explain their desirable and undesirable attributes. Depending on the conservation objective of the government, those farmers contributing to that objective can, accordingly, be awarded. In the long-run, those selected farmers can be targeted for future on-farm conservation.

- *Benefit sharing arrangements*: In this case, public crop diversity conservators should negotiate with users of genetic information maintained by farmers (research organizations, seed companies and pharmaceuticals) so that farmers can share part of the benefit derived. This option is more appropriate for those genetic resources that can result in marketable products like IVs, medicine for humans and animals, and biotechnology products. This, however, requires creating adequate demand for indigenous varieties by potential genetic information users and resolving problems presented in Section 4.3.

Building upon the household level analysis conducted, this chapter has estimated national level opportunity costs and presented institutional incentive schemes for undertaking on-farm conservation. The topics covered in this chapter and the conclusions derived are complementary to the concluding chapter that follows.

9. Summary, conclusions, and policy implications

9.1. Summary: recapping matters of cross-cutting importance

This thesis has been motivated by the fact that, among the available conservation options, on-farm conservation of crop diversity is both appealing and challenging with many issues left still open and under discussion. Many questions demand critical examination: How should on-farm conservation be undertaken? Which farmers should be targeted? How much farmers should be compensated? How much should be invested? And what institutions should prevail?

On-farm conservation deserves further study for different reasons. First and foremost, farmers are managing the huge stock of crop diversity. Secondly, so far there are no established principles to make it operational. Despite the existence of *de facto* conservation that is as old as farming, on-farm conservation strategy (as public conservation strategy) is yet emerging. On-farm conservation involves selecting varieties to be conserved, identifying the agro-ecologically representative sites across localities, and negotiating with few selected farmers in the respective locations. How can this be operational?

Policy can be more successful in influencing variety choice behavior of farm households if their diversification motives, variety attribute preferences and the opportunity costs involved are well understood. The thesis has dwelled on these topics taking coffee, sorghum, and wheat data as examples for empirical illustrations. The results are meant to be inputs to identify farmers and varieties to be targeted. They are also meant to serve as inputs to design sensible contextual policy incentives for on-farm conservation, variety development and diffusion.

9.2. Major conclusions and policy implications

The thesis results have implications not only for on-farm conservation but also for improved variety adoption and breeding priority setting. Since the use of crop genetic resources has to be part of the conservation exercise, this chapter will pinpoint the relevance of the results to both of these policy concerns.

9.2.1. Conclusions

The results have shown that, as such, adoption of improved varieties is not an immediate threat to crop diversity in Ethiopia. This is evident as most farmers are not replacing local varieties by improved varieties but using both side by side. Due to failure of IVs to satisfy farmers' multiple concerns, automatic replacement is unlikely to happen. However, lower crop diversity on farmers' fields with more experience in using IVs suggests that new varieties can threaten crop diversity over the longer term.

The results further imply that more risky (in terms of frequent occurrence of diseases and pests) and more diverse farming systems are less vulnerable to crop uniformity. Accordingly, maintaining a diverse set of varieties is mainly a poor, subsistent and marginalized farmers' undertaking with limited access to markets, roads, and extension. These farmers seem to have recognized various benefits of poly-variety - risk reduction, food tastes, tolerance to environmental stress (weather, pests, and weed), addressing their multiple concerns, and *asset specificity* benefits.

Access to markets and extension detaches the link between *de facto* conservation and farmers' '*survival first*' strategies. In the case of sorghum, cash crop (cht'at) farming is found to have a negative effect on its diversity. Farmers' multiple objectives and concerns, risk and yield insurance considerations, land heterogeneity, labor endowment, and lack of access to markets have been the factors inducing the production of sorghum diversity.

These results lead us to conclude that designing an incentive scheme with subsistence and marginalized farmers can be done more easily with less cost implications. These farmers are found to have higher propensity to grow local varieties. Similarly, designing policy incentives with farmers having multiple concerns, larger land holding, and better experience in farming is relatively easy. On the contrary, it is either difficult or costly to negotiate and design policy incentives with households characterized by better access to markets, agricultural extension, input supply, and cash crop farming possibilities.

As the relative importance of coffee and sorghum decreases, the probability of losing their diversity gets higher. Conserving the diversity of less rewarding food crops (like sorghum), in an area where planting a highly rewarding cash crops (like cht'at) is also an option for farmers, is either difficult or costly. This is mainly because opportunity cost compensation could go far more than the net-return farmers expect from growing the food crop. Entrusting these farmers can increase the *moral hazard* problem unless the incentive satisfies their expectations considering the return they derive from the more rewarding crop.

As rural development interventions (improved access to extension, IVs, markets, and growing cash crops) are put in place, the land allocated to FVs and the amount of varieties per farm and, therefore, the level of on-farm crop diversity produced will shrink. Thus, 'development' will come about at the expense of crop diversity leading to uniformity in crop varieties.

In areas where the in-built incentives are prevailing, attempting to address the diverse concerns of farmers with a single and genetically uniform improved variety is unlikely to be acceptable to farmers.

The variety attribute preference analysis results suggest that farmers in more accessible areas and those who are less concerned with satisfying subsistence income prefer yield and marketability. On the contrary, farmers in less accessible areas (for markets and extension services) and those who are more concerned with potential future income shocks have more propensities for adaptability and yield stability. Thus, high yielding and marketable varieties will be conserved *de facto* by the first group of farmers and adaptable and stable varieties will be conserved *de facto* by the second group.

As land and labor get scarce, farmers' demand for yield increases. In contrast, a decrease in the opportunity cost of land and labor will increase farmers' demand for yield stability. Moreover, farmers who are less concerned with natural problems (disease, drought, and pests) have been found to have higher demand for yield. Similarly, farmers who expect good prospect for availability of rural development services (input supply, farm implements, markets and extension services) have higher demand for yield and marketability.

Those farmers who have given better rank to the local varieties with respect to yield and yield stability have been found to plant higher proportion of local coffee varieties on-farm. Thus, coffee varieties embedding yield and yield stability attributes will be conserved *de facto* in most coffee growing areas. On the contrary, farmers who have given better rank to the local varieties with respect to marketability have reported lower proportion of local coffee trees on-farm. These farmers are trying to get other non-marketability attributes from improved varieties.

The opportunity cost results have shown that (for both sorghum and wheat) access to market and extension package, marketing surplus, fertilizer use, and experience in growing IVs are the most important variables increasing opportunity costs. On the contrary, land quality, input prices, education level of the household head, and oxen ownership reduce the opportunity costs. Thus, IVs do not necessarily have better comparative advantages with better quality plots. FVs can even have better comparative advantage with good quality plots. Similarly, better-educated and better endowed (with oxen) farmers do not necessarily have comparative advantage in using IVs. High input prices hit most the UIVs reflecting the capital intensity of planting them. By way of comparing opportunity costs for sorghum and wheat, the negative impact of input prices and the positive impact of output prices were found to be stronger for wheat.

IVs are found to have comparative advantage when accompanied by other inputs like extension, fertilizer, credit, markets and storage facilities. Moreover, the results have shown that asset ownership like oxen increases the comparative advantage of using FVs. Farmers who can apply fertilizer and labor easily have also better comparative advantage with IVs. So far, Ethiopian smallholders have not been able to fully benefit from inputs like IVs because of capital limitations, nature dependence of agriculture, and partial adoption of the technologies being promoted.

Farmers have better comparative advantage of using IVs if the output is mainly produced for the market. Moreover, as the relative importance of sorghum to the livelihood of a household decreases, the comparative advantage of using IVs also decreases. Sorghum growing farmers earning the majority of their income from cash crop farming (cht'at) have lower comparative advantage of growing IVs.

The national level *ex ante* on-farm conservation cost estimates (for sorghum, wheat, and coffee) reveal that as the crop becomes more valuable in the market, the unit cost of maintaining that landrace on-farm increases. For instance, based on the opportunity cost estimates, maintaining one landrace of sorghum would cost the government between 324 Birr (€ 32) and 1,302 Birr (€ 130) annually while maintaining one landrace of wheat would cost between 830 Birr (€ 83) and 1,306 Birr (€ 130). For coffee, this estimate, based on farmers' willingness to accept for coffee land is 1038 Birr (€ 104) to 2409 Birr (€ 241). Coming to the gene bank conservation cost estimates, the operating cost per accession for the years 1990 to 2001 ranges from 15 to 56 Birr. However, on average, only 36 to 52 percent of its capacity is utilized during those years. If the country manages to utilize its gene bank at its full capacity, there is a huge potential to reduce the unit cost of maintaining an accession by about 92 to 179 percent.

9.2.2. Policy recommendations

Before implementing any of the policy recommendations provided below, government should identify endangered species and hot-spots, crops and varieties to be maintained by policy, and those conserved *de facto*. Following this, it has to set the type and level of effort expected from farmers. Once hot-spots and endangered locations are identified, it is a question of negotiating and placing contextual policy incentives. It has been argued in the thesis that this has to be done building upon farmers' incentives, attribute preferences, and opportunity costs. This should enable them to act according to the policy requirements.

Since on-farm conservation calls for dealing with few farmers to reduce the national level opportunity costs to be foregone, its efficient implementation requires targeting principles. The targeting principles could be set to target either pro-diversity farmers and/or farmers living in areas where less crop diversity is produced. While the motive to target the first group of farmers emanates from their willingness to produce diversity, targeting the second group is motivated by the desire to salvage endangered species.

To target the first group of households, policy makers have to approach those farm households protecting their livelihoods by diversifying on variety portfolio. To this end, conservation measures should target farmers who have higher in-built incentive to maintain and diversify on local varieties of crops. On the contrary, targeting the second group calls for approaching localities and farmers where the in-built dis-incentives prevail. In this scenario, on-farm conservation can be more costly as it will be hard for these farmers to switch to varieties of policy interest. In areas where there is meager *de facto* conservation, government can better undertake conservation by itself to reduce *moral hazard* problem after the contract. Moreover, from the point of view of farmers, the rest of the society is the residual claimant of the benefits of conservation.

In a multi-crop farming system where crops have different relative importance to farm households, unique varieties of the less important crop(s) sufficiently important for conservation should get prior conservation actions to salvage their potential loss.

For farmers residing in areas where in-built incentives are prevailing, breeders should increase their choice set so that they can take the varieties that fit to their contextual working environment because no single variety can satisfy all these farmers' household needs. Hence, breeding strategies should aim at diverse variety attributes, not just yield or disease resistance. Accordingly, genetically uniform IVs can rather be targeted to more accessible areas with better experience in IVs and less vulnerability to potential income shocks.

When marketable and high yielding local varieties of coffee are planted by the better accessible and income-shock tolerant farmers, there is no need to design any incentive in these areas to maintain these varieties. Similarly, stable and adaptable varieties in the hands of inaccessible and income-shock vulnerable farm households do not need any policy incentive.

In an area where land and labor are relatively scarce, compensation for on-farm conservation should be based on yield and marketability. The same compensation strategy should be followed with better-off farmers who have better access to input supply, irrigation, extension services, and farm implements. In such localities, on-farm conservation should target varieties that do not satisfy yield and marketability attributes as they are not conserved *de facto*. To address variety demands of farmers with the above features and those operating in irrigated areas, the most important variety attributes worth considering are yield and marketability.

On the contrary, yield stability and environmental adaptability should be used as a reference for farmers who fail to satisfy their current consumption requirement located in less accessible areas. Conservation in such cases should target varieties that do not satisfy yield stability and environmental adaptability. To address variety demands of such farm households, the priority variety attributes are yield stability and environmental adaptation. For annual crops like sorghum, early maturity is also very important variety attribute to consider.

With the government's limited capacity to satisfy demand for multiple variety attributes, labor constrained, experienced (with IVs), and better-educated farmers are the ones to be targeted for variety adoption with single most important attribute like yield or marketability. On the contrary, farmers with more experience in growing local varieties of coffee or better land holding or those with no access to markets are the ones worth considering for distribution of variety (ies) with multiple attributes.

Given that farmers' variety attribute preferences are determining both their propensity to use IVs and the chance of using them productively, breeding should target to satisfy demands of different farm household types classified by their resource endowments, preferences, and constraints. The research priority setting should, therefore, ask '*breeding for whom?*' not only just '*breeding for which environment?*', as this could make the breeding program more responsive to smallholders' demands.

Farm households found in localities where factors increasing opportunity costs prevail should get better compensation and incentives should, in principle, be made flexible depending on the variability of opportunity costs temporally and spatially. Optimizing on-farm conservation costs calls for targeting farmers and areas where IVs have no comparative advantage.

In undertaking on-farm conservation in both accessible and inaccessible areas, the compensation will be higher for farmers of accessible areas. If the marginal farmers remain inaccessible, the outcome will marginalize the poorest of the poor. Lower compensation for these farmers should generate an incentive for the government to improve access differences in the long-run. Such a policy action is motivated by the need for reducing relative deprivation. Heterogeneous compensation improves compliance because it satisfies farmers' contextual expectations and is cost effective because the payment is based on opportunity costs. Of course, it can complicate monitoring and enforcement especially in the short-run.

On-farm conservation with farmers who are paying higher opportunity costs is more costly as it will be harder to convince them to switch to variety (ies) selected for maintenance by policy makers. Similarly, in a farming system where high value cash crops (like cht'at, coffee and vegetables) prevail, incentives (for maintaining local varieties of food crops like sorghum and wheat) should be crafted based on the benefits foregone from high value crops.

Results of the opportunity cost analysis also imply that recommending IVs across the board for all farmers is a waste of resources and FVs can work even better in some farmers' working environments. Equally destructive is the notion that IVs do not work in less-favored areas.

Mostly, it is argued that as long as the subsistence farmers have no opportunity to get hold of the accessories, they do not have comparative advantage in using IVs. But improving access variables can reverse the argument because policy variables can be improved to realize the comparative advantage. By virtue of being in areas where crop diversity prevails, pro-diversity farmers should not shoulder their service of maintaining crop diversity for free being deprived of using better varieties and opportunities.

Targeting adoption to only accessible farmers further deprives poor farmers. Thus, targeting the so called '*high potential areas and farmers*' for distribution of IVs has to be re-visited since all of these farmers do not have comparative advantage in using IVs. Similarly, all of the farmers in less-favored areas do not have comparative dis-advantage. As the average productivities of IVs in '*high potential*' versus '*less-favored areas*' do not tell us the complete story, the current extension strategy has to target the '*less-favored areas*' too.

9.3. Future research needs

There is still a lot unknown in the areas of what on-farm conservation entails and what it renders in different situations. As it is still emerging, experiences are scanty and research gaps are vast. With this point in mind, the following research agenda are suggested for future work.

As an extension to the empirical analysis done in Chapter 5, the following research areas are suggested. The impact of crop diversity on yield stability (risk reduction) in a panel data setting is one important area of research worth undertaking. The role of on-farm crop diversity on farmers' efficiency is another important research area. Equally important is the analysis of the impact of crop diversity on farmers' livelihoods – poverty, food security and productivity. Studies on the human dimensions of on-farm conservation are non-existent. There is also huge potential to study the linkage between on-farm conservation, crop insurance and agricultural marketing.

In theory, the variety that one farmer reports as a single variety could be genetically more or less diverse than a variety planted and reported by another farmer. Even though the linkage (between crop genetic diversity and farmers' decisions) is hard to establish, monitoring farmers' cultural practices and relating those to the outcome variable can shade more light on the dynamics of farmers' variety management.

For a descriptive choice model (like MNL) to be made operational, the alternatives must be mutually exclusive, the choice set must be exhaustive, and the number of alternatives must be finite (Train, 2003). Even though different in some respects, attributes like disease resistance, yield stability and environmental adaptability are inter-connected. The attribute preference analysis in Chapter 6 can be extended in such a way that the variety attribute choice is made mutually exclusive. Combining all possible attribute combinations of interest and presenting them to farmers as mutually exclusive options can better elicit their revealed preferences. Moreover, offering more concise definitions of yield levels and market prices to farmers can make the choice exercise more precise.

Studying the marginal substitution among the different variety attributes using a choice experiment framework is another interesting area of investigation. For instance, what does early maturity mean in terms of yield stability or food taste or yield? Translating non-monetary variety attributes in monetary terms can provide a complete picture regarding the relative importance of the different variety attributes and full extent of opportunity costs foregone during a change in variety use. This way, one can understand the trade-offs farmers are prepared to face as far as variety attributes are concerned.

Estimating on-farm conservation costs based on *ex post* on-farm conservation expenditures can provide a chance not only to compare these costs with ex situ conservation costs (if the outcomes are similar) but also to simulate the national level opportunity costs of undertaking on-farm conservation.

Appendices

Appendix I. Glossary of some terms used in the thesis[1]

Accession is a sample of a crop variety collected at a specific location and time (Heywood, 1995).

Different inputs are suited to different crop enterprises and varieties differently. *Asset specificity* refers to lack of suitability of an asset to all enterprises or crop varieties equally.

Biological diversity refers to the variability among living organisms from all sources including, *inter alia*, terrestrial, marine and other aquatic ecosystems and the ecological complexes of which they are part (Heywood, 1995). It can be considered at three levels *i.e.* diversity within species (*genetic diversity*), between species (*species diversity*) and between ecosystems *(ecosystem diversity)* (Weitzman, 1993). In this study, species diversity is the point of interest and it is proxied by the number of local varieties used by farmers as identified by themselves or by the number of desirable local variety attributes.

Common-pool resources (CPRs) are defined as a class of resources for which exclusion (excludability) is difficult and joint use involves subtractability (rivalry) (Berkes *et.al*, 1989). In this thesis, crop diversity is considered in a CPR framework because these resources are non-excludable, gifts of nature owned by the community, and their joint use involves conservation.

Conservation is the management of human use of the biosphere so that it may yield the greatest sustainable benefit to present generations while maintaining its potential to meet the needs and aspirations of future generations (IUCN, 1980). It is important to note that 'conservation' does not mean non-use, but rather 'wise' use that contributes to sustainable development (McNeely, 1988). Of course, conservation can not rely solely on economic and ecological justifications (Ehrenfeld, 1976).

Crop diversity refers to resources including farmers' varieties (cultivated species), and non-cultivated species from other plant species or even species outside the plant kingdom (Heywood, 1995). It is the information in

[1]The term defined is in italics.

the genetic resource, not so much the tangible resource itself that is valuable in the research and development process (Swanson and Göschl, 2000b). Crop diversity on-farm in this thesis is designated by the number of local varieties on farmers' fields, number of variety attributes important to the farm household, and the relative importance of local varieties on farmers' fields.

De facto conservation is the decision of farmers to continue cultivating local varieties (Meng, 1997). It is the positive externality of farmers' attempt to satisfy their household concerns. *De facto* conservation, in this study, refers to level of crop diversity that farmers maintain (as a positive externality of their survival first strategies) while trying to achieve their livelihood objectives. It is the inadvertent result of farmers' derived demand for crop diversity.

Direct use value refers to economic values derived from direct use or interaction with a biological resource (Heywood, 1995).

Ex situ conservation literally means out of place – not in the original or natural environment (IBPGR, 1991). It is a conservation strategy that involves keeping components of crop diversity alive outside their natural habitats or natural environment, for instance, by using zoos, botanical gardens, and gene banks (Heywood, 1995).

Farmers' varieties (FVs) refer to crop varieties that have been maintained by farmers and have evolved on farmers' fields during their farming experience (This thesis). The phrases farmers' varieties, local varieties, and indigenous varieties are used interchangeably throughout the thesis.

Gene bank is a storage facility where germplasm is stored in the form of seeds, pollen, embryos, or pollen (Heywood, 1995).

Genetic diversity is a complex statistical concept referring to the variance among alleles at individual gene loci, between individuals, within populations, and between populations (Brown, 1983).

Genetic material means any material of plant, animal, microbial or other origin containing functional units of heredity (Heywood, 1995).

Germplasm may be defined as seeds, plants or plant parts that are useful in crop breeding, research or conservation because of their genetic attributes (Frisvold and Condon, 1998).

Improved varieties (IVs) refer to crop varieties developed by modern plant breeders designed to address a diverse set of production and consumption related problems. They are varieties of crops that come from breeding either as hybrids or just selections from the collections taken from farmers' fields (This thesis).

In- built dis-incentives are household related contextual factors that detach the link between crop diversity and farmers' survival strategies (This thesis).

Impure public goods are goods that are neither purely public not private but produced as a positive externality of activities of private economic agents (Cornes and Sandler, 1986). Example: *De facto* conservation of crop diversity by farmers.

In-built incentives are farmers' incentives to produce crop diversity derived from within their household concerns, endowments and constraints (This thesis).

Indirect use or derived value refers to the economic value derived from the role of resources in supporting activities whose outputs have direct value in production or consumption (Heywood, 1995).

In situ conservation involves conserving crop diversity using traditional agriculture and nature as a media. It is the conservation of crop diversity within the evolutionary dynamic ecosystems of the original habitat or natural environment (Heywood, 1995). *In situ* conservation, in this thesis, refers to preserving varieties cultivated by farmers using their own selection methods and criteria (FAO, 1989; Bommer, 1991).

Insurance value refers to the value of crop diversity in maintaining and stabilizing the productivity of cereal production (This thesis).

Intellectual property right is the right enabling an inventor to exclude imitators from the market for a limited time (Heywood, 1995).

Landrace is a plant population with a limited range of genetic variation that is adapted to local agro-climatic conditions and which has been generated, selected, named and maintained by traditional farmers in response to a specific physical environment and specific social or economic constraints (Teshome *et.al*, 1999).

On-farm conservation refers to the sustainable management of genetic diversity of locally developed indigenous crop varieties, with associated wild and weedy species or forms, by farmers within traditional agricultural systems (Maxted *et.al*, 1997). Conservation on-farm is the domain of in situ conservation of crop diversity (UNEP, 1992).

Opportunity cost refers to loss of potential earnings when farmers are expected to plant local varieties of crops identified by policy for conservation purpose (This thesis).

Policy incentives refer to specific policy inducements meant to influence farmers' behavior so that they can make pro-diversity decisions (This thesis).

Perverse incentives are policy actions that slow down the production of crop diversity (This thesis). Example: Subsidies on IVs.

Poly-variety refers to planting multiple varieties of a single crop in the same field (Netting and Stone, 1996).

Recalcitrant seed is a seed that does not survive drying and freezing (Heywood, 1995).

Resilience refers to the capacity of a system to absorb shocks without losing stability (Holling, 1973).

Revealed preference approaches are approaches that include derived demand approaches and infer from data on behavioral changes in actual markets related in some way to the missing market(s) (Heywood, 1995).

Seed bank is a facility designed for the ex situ conservation of individual plant samples through seed preservation and storage (Heywood, 1995).

Variety is a type of crop (say sorghum) that is named by farmers having certain phenotypic characteristics that enable them to identify one from the other (This thesis).

Appendix II. Brief description of two regression models

Appendix 2-A Ordered probit regression model

The farmer's utility function reflects subjective expectations that are determined by household level contextual characteristics. Therefore, farmers' utility associated with variety attribute (s) or the number of local varieties they are managing is given as:

$$y_i = \beta' x_i + u_i$$

where y_i is the utility level attained by the farmer with choice i (1 if the farmer is using only one variety or giving rank 1 to a certain variety attribute, 2 if he is using 2 varieties or giving rank 2 to a certain variety attribute and 3 if the on-farm count diversity is 3 or the farmer is giving rank 3 to a certain variety attribute and so on). In this setting, both on-farm crop diversity and the rank of each variety attribute are ordered.

The ordered probit model assumes that there are cut-off points - Y^* and Y^{**} - that define the relationship between the observed and unobserved dependent variables (Pindyck and Rubinfeld, 1998). In the sample, there are farmers having a count index value of 0, 1, and 2[2]; farmers' ranking of variety attributes (1, 2, 3....). Let us denote the first group of farmers, respectively, as D_i^0, D_i^1, and D_i^2. The notations D_i^0, D_i^1, and D_i^2 get 1 if the farmer has a count diversity of 0, 1 and 2, and 0 otherwise, respectively. The ordered probit model (Zavoina and McElvey, 1975) could, thus, be presented as:

$$D_i^0 = 1 \, if \; x_i\beta + u_i < Y^*$$
$$D_i^1 = 1 \, if \; Y^* < x_i\beta + u_i < Y^{**}$$
$$D_i^2 = 1 \, if \; x_i\beta + u_i > Y^{**}$$

where the x_i are the explanatory variables considered in Tables IV-1, IV-5 and IV-8, β is a vector of unknown parameters to be estimated and u_i are the independently and identically distributed error terms with zero mean and constant variance, σ^2.

Assuming that the error terms (u_i) are uncorrelated, it follows that:

$\Pr(D_i^0 = 1) = \Pr(Y_i = 0) = \Pr(x_i\beta \leq Y^*) = \Phi(Y^* - x_i\beta)$,

$\Pr(D_i^1 = 1) = \Pr(Y_i = 1) = \Pr(Y^* < x_i\beta < Y^{**}) = \Phi(Y^{**} - x_i\beta) - \Phi(Y^* - x_i\beta)$,

$\Pr(D_i^2 = 1) = \Pr(Y_i = 2) = \Pr(x_i\beta \geq Y^{**}) = 1 - \Pr(D_i^0 = 1) - \Pr(D_i^1 = 1) = 1 - \Phi(Y^{**} - x_i\beta)$[3].

[2]It has to be noted, however, that specific values assigned to the dependent variable are not relevant in ordered probit. Only the ordering implied by the values is what matters. Thus, for instance, 0, 1, 2 is equally treated as 3, 475 and 597.

[3]This holds true because the probability distribution of the probit is symmetric.

where Φ is the cumulative standard normal distribution (*CDF*).

The likelihood function is the product of the individual probabilities taken over-all 'n' observations. Denoting that 1 ,..., n_1 as the first set of observations such that $Y_i = 0$, $n_1+1,\ldots, n_2$ as the second set of observations such that $Y_i = 1$ and $n_2+1, \ldots, n_3$ as the third set of $n - n_1 - n_2$ observations such that $Y_i = 2$, the likelihood function will be:

$$L = \underbrace{\Pr(Y_1 = 0) * \Pr(Y_2 = 0) \ldots \Pr(Y_{n_1} = 0)}_{\prod_{i=1}^{n_1}[F(Y^{*} - x_i\beta)]} * \underbrace{\Pr(Y_{n_1} + 1 = 1) \ldots \Pr(Y_{n_2} = 1)}_{\prod_{i=n_1+1}^{n_2}[F(Y^{**} - x_i\beta) - F(Y^{*} - x_i\beta)]} * \underbrace{\Pr(Y_{n_2} + 1 = 2) \ldots \Pr(Y_n = 2)}_{\prod_{i=n_2+1}^{n}[1 - F(Y^{**} - x_i\beta)]}.$$

Suppose the function to be estimated is: $\alpha + \beta x_i$, dividing this by σ for the purpose of normalization,

$$\Pr(D_i^0 = 1) = \Phi\left(\frac{Y^{**} - \alpha - \beta x_i}{\sigma}\right)$$

$$\Pr(D_i^1 = 1) = \Phi\left(\frac{Y^{**} - \alpha - \beta x_i}{\sigma}\right) - \Phi\left(\frac{Y^{*} - \alpha - \beta x_i}{\sigma}\right)$$

$$\Pr(D_i^2 = 1) = 1 - \Phi\left(\frac{Y^{**} - \alpha - \beta x_i}{\sigma}\right)$$

When there are only three choices, Y^* can be set equal to zero without loss of generality (Johnson and DiNardo, 1997). Thus, the above likelihood function can be given as:

$$L = \prod_{i=1}^{n_1}[\Phi(-x_i\beta)] * \prod_{i=n_1+1}^{n_2}[\Phi(Y^{**} - x_i\beta) - \Phi(-x_i\beta)] * \prod_{i=n_2+1}^{n}[1 - \Phi(Y^{**} - x_i\beta)]$$

$$= \Phi\left(\frac{Y^{**} - \alpha - \beta x_i}{\sigma}\right) . * \left[\Phi\left(\frac{Y^{**} - \alpha - \beta x_i}{\sigma}\right) - \Phi\left(\frac{-\alpha - \beta x_i}{\sigma}\right)\right] * \left[1 - \Phi\left(\frac{Y^{**} - \alpha - \beta x_i}{\sigma}\right)\right]$$

The treshold value Y^{**} is estimated along with the β coefficients by maximizing the following log-likelihood function:

$$L_{oprobit} = \log\Phi\left(\frac{Y^{**} - \alpha - \beta x_i}{\sigma}\right) + \log\left[\Phi\left(\frac{Y^{**} - \alpha - \beta x_i}{\sigma}\right) - \Phi\left(\frac{0 - \alpha - \beta x_i}{\sigma}\right)\right] + \log\left[1 - \Phi\left(\frac{Y^{**} - \alpha - \beta x_i}{\sigma}\right)\right]$$

For the three probabilities, the marginal effects of the change in the regressors are:

$$\frac{\partial \Pr(Y_i = 0)}{\partial x_k} = -\phi(-\alpha - \beta x_i)\beta,$$

$$\frac{\partial \Pr(Y_i = 1)}{\partial x_k} = -[\phi(Y^{**} - \alpha - \beta x_i) - \phi(-\alpha - \beta x_i)]\beta$$

$$\frac{\partial \Pr(Y_i = 2)}{\partial x_k} = \phi(Y^{**} - \alpha - \beta x_i)\beta$$

where ϕ is the probability density function (*PDF*).

The interpretation of the coefficients is in terms of the underlying latent variable model. For example, a positive β_k means that the corresponding variable increases count diversity. In other words, the latent variable Y_i^* increases if x_{ik} increases. Accordingly, the probability that $Y_i = 2$ will increase, while the probability that $Y_i = 0$ will decrease. The effect on the intermediate categories is, however, ambiguous (Verbeek, 2000).

Appendix 2-B Two-limit Tobit

As far as the effect of use of IVs on crop diversity is concerned, the intensity of use is more important than the discrete choice decision. The Tobit model measures not only the probability that a farmer will use a new technology but also the intensity of the use of the technology once adopted.

The stochastic model underlying Tobit may be expressed by the following relationship:

$$Y_i = x_i\beta + \varepsilon_i \text{ ; if } x_i\beta + \varepsilon_i > 0$$

$$= 0 \text{ ; if } x_i\beta + \varepsilon_i < 0.$$

where Y_i refers to the censored variable (in our case variables like ShaNNon, SHANEVEN, IMPROPOR, and PROPTRES), x_i is the vector of the regressors, β is a vector of the parameters to be estimated, and ε_i is an independently distributed error term assumed to be normal with zero mean and constant variance, σ^2.

The loglikelihood function for the Tobit is:

$$l_{Tobit} = \underbrace{\sum_{y_i|y_i=0} \ln\left[1 - \Phi\left(\frac{x_i\beta}{\sigma}\right)\right]}_{P(y_i \leq 0)} + \underbrace{\sum_{y_i|y_i>0}\left[\ln\frac{1}{\sqrt{2\pi\sigma^2}} - \frac{1}{2}\frac{(y_i - x_i\beta)^2}{\sigma^2}\right]}_{P(y_i|y_i)}$$

Given x_i, the first part is the probability that $y_i = 0$. The second part is the distribution of y_i given that it is positive. Amemiya (1984) has shown that the Tobit maximum likelihood estimator remains consistent under serial correlation but not under heteroscedasticity or non-normality. For this reason, the Tobit models are estimated with correction for heteroscedasticity and without correction for heteroscedasticity. In all the cases, the likelihood ratio test supports the Tobit results with correction for heteroscedasticity. All the Tobit results presented in the Appendix are, therefore, Stata's robust estimation results.

The Tobit model has an advantage in that its coefficients can be further disaggregated to determine the effect of a change in the ith variable on changes in the probability of keeping more than one indigenous variety and the expected intensity of use of the variety (McDonald and Moffit, 1980). The decomposition is given as:

$$\frac{\partial E[y_i \mid x_i]}{\partial x_k} = \left\{ \underbrace{\Phi\left(\frac{x_i\beta}{\sigma}\right)}_{prob(y_i^*>0)} \right\} \underbrace{\beta_k \left[1 - \frac{x_i\beta}{\sigma} \frac{\phi\left(\frac{x_i\beta}{\sigma}\right)}{\Phi\left(\frac{x_i\beta}{\sigma}\right)} - \left(\frac{\phi\left(\frac{x_i\beta}{\sigma}\right)}{\Phi\left(\frac{x_i\beta}{\sigma}\right)} \right)^2 \right]}_{\frac{\partial E[y_i \mid x_i, y_i^*]}{\partial x_k}} + \underbrace{\left\{ \frac{\beta_k}{\sigma} \phi\left(\frac{x_i\beta}{\sigma}\right) \cdot \left(x_i\beta + \frac{\phi\left(\frac{x_i\beta}{\sigma}\right)}{\Phi\left(\frac{x_i\beta}{\sigma}\right)} \right) \right\}}_{\frac{\partial prob(y_i^*>0)}{\partial x_k} E[y_i \mid x_i, y_i^*>0]}$$

One effect works by changing the conditional mean of the diversity index, given that it is observed (greater than zero) and the second by changing the probability that an index will be positive, the second part (Johnston and DiNardo, 1997). While the first is about the effect of a unit change in any of the regressor on the level of diversity (once it is positive), the second is about the effect of a unit change in any of the regressor on the probability of having a positive level of crop diversity.

As our dependent variables are not taking values beyond 0 and 1, the two-limit Tobit is employed for the results reported in Tables IV-3, IV-4, IV-7, and IV-9.

Appendix III. Food consumption equivalence scale and additional descriptive statistics

Table III-1: Equivalence scale of recommended energy intakes by age categories (employed in Chapters 5 and 6)

Category	Age (Years)	Average energy allowance per day (kilo calorie)	Equivalence scale
Infants and	0 – 0.5	650	0.22
children	0.5-1	850	0.29
	1-3	1300	0.45
	4-6	1800	0.62
	7-10	2000	0.69
Males	11-14	2500	0.86
	15-18	3000	1.03
	19-25	2900	1
	25-50	2900	1
	51+	2300	0.79
Females	11-14	2200	0.76
	15-18	2200	0.76
	19-25	2200	0.76
	25-50	2200	0.76
	51+	1900	0.66

Source: NRS (1989)

Table III-2: Sorghum farmers' valuation of land (in Birr per Hectare) (employed in Chapters 7 and 8)

Pas[4]	Mean value of willingness to pay for renting in	Mean value of willingness to accept for renting out
Belina Arba (N = 15)	386.0	399.6
Umerkule (N = 11)	321.7	381.4
Kerodeda (N = 20)	510.5	523.0
Asseliso (N = 18)	679.2	706.9
Eje Aneni (N = 27)	448.2	456.0
Chachole (N = 22)	325.0	343.2
Gurbo (N = 12)	366.7	366.7
Total (N = 123)	454.2	468.8

[4]Figures in parentheses refer to the number of respondents in each PA.

Table III-3: Coffee farmers' valuation of land (Birr per Fechasas[5]) (employed in Chapters 7 and 8)

PAs	Mean value of willingness to pay for renting in	Mean value of willingness to accept for renting out
Harro (N = 30)	143.9	157.3
Kella Guda (N = 27)	187.7	200.6
Bulbullo (N = 22)	155.8	167.4
Kilole Kirkir (N = 28)	123.0	133.8
Yachi Urechi (N = 28)	117.9	130.9
Gibe Boso (N = 11)	156.3	169.1
Hallo Sebeka (N = 24)	127.8	136.9
Sebeka Debiye (N = 23)	78.5	86.5
Total (N = 192)	135.6	143.9

[5]Fechasa is the most common unit of measurement for land in the Jimma area. 1 Hectare = 4 Fechasa.

Table III-4: Frequency of the count and discrete variables (employed in Chapter 5)

Variable	Description	Value	Freq.	Percent	Cum. Freq.
COUNT	See taxt	0	75	38.1	38.1
		1	84	42.6	80.7
		2	38	19.3	100.0
AGRODIV	Total number of annual and perennial crops less one	0	34	17.2	17.2
		1	75	37.9	55.1
		2	59	29.8	84.8
		3	30	15.2	100.0
NONFFARM	See text	0	114	58.8	58.8
		1	80	41.2	100.0
		1	158	79.8	100.0
NOPURPOS	See text	1	56	28.3	28.3
		2	50	25.3	53.5
		3	37	18.7	72.2
		4	41	20.7	92.9
		5	12	6.1	99.0
		6	2	1.0	100.0
CREDIT	See text	0	124	64.6	64.6
		1	68	35.4	100.0
CHT'AT	See text	0	82	41.4	41.4
		1	31	15.7	57.1
		2	46	23.2	80.3
		3	39	19.7	100.0
PLOTS	See text	1	83	43.9	43.9
		2	80	42.3	86.2
		3	20	10.6	96.8
		4	3	1.6	98.4
		5	3	1.6	100
VISITS		0	100	50.5	50.5
		1	22	11.1	61.6
		2	23	11.6	73.2
		3	15	7.6	80.8
		4	8	4.0	84.8
		5-12	30	15.17	100

Table III-5: Description of clustered variables (employed in Chapter 5)

Variable	Description	No. of zeros	Mean	SD	Min.	Max.
IMPROPOR	Proportion of sorghum land allocated for the improved variety as a percentage of all sorghum land	122	0.192	0.290	0	1
ShaNNon	ShaNNon diversity index	76	0.487	0.420	0	1.39
SHANEVEN	ShaNNon's evenness diversity index	76	0.598	0.456	0	1

Appendix IV. Additional regression results

Appendix 4-A . Supplement to Chapter 5[6]

Table IV-1: Ordered probit regression results to explain farmers' motivations to diversify on local sorghum varieties

Variable	Coefficient	Marginal effects: Dy/Dx Count = 0	Count = 1	Count = 2
Age	0.01792* (1.84)	-0.005	0.0037	0.0013
Cht'at	-0.3748*** (-3.26)	0.105	-0.078	-0.027
Lancrops	0.52732*** (4.06)	-0.148	0.110	0.038
Impexep	-0.4013*** (-5.22)	0.113	-0.084	-0.029
Nopurpos	0.7543*** (6.76)	-0.212	0.157	0.054
Totlabor	0.090* (1.69)	-0.025	0.019	0.0065
Access	0.0091** (2.09)	-0.003	0.0019	0.0007
Educate	0.0554 (1.00)	-0.016	0.012	0.004
Plots	0.2902* (1.86)	-0.081	0.061	0.021
$Risk_{proxy}$	-0.0008 (-2.89)	0.0002	-0.0002	-0.00006
Belinarba*	0.2851 (0.73)	-0.073	0.048	0.0248
Kerodeda*	1.052** (2.48)	-0.212	0.070	0.142
Asseliso*	1.554*** (3.37)	-0.270	0.005	0.265
Ejeaneni*	1.668*** (3.63)	-0.278	-0.023	0.301
Chachole*	-0.2291 (-0.43)	0.068	-0.05	-0.014
Gurbo*	-0.1065 (-0.16)	0.031	-0.02	-0.007
Credit*	0.1995 (0.80)	-0.056	0.042	0.014
Nonffarm*	0.0307 (0.13)	-0.009	0.006	0.002
Constant	-3.46 (-5.29)			
Dependent variable is count		Number of obs = 178		
Wald χ^2 (18) = 157.63		Prob (χ^2) = 0.00		
Loglikelihood = -99.19		Pseudo R^2 = 0.468		

Notes: ***-Significant at 1%; **- Significant at 5%; and *- Significant at 10%. Values in parentheses are the ratio of the coefficient to the estimated asymptotic standard error. The method of estimation is Stata's Robust option following Huber/White standard errors and covariance. (*) Dy/Dx is for discrete change of dummy variable from 0 to 1

[6] The description of the response variables of Tables IV-3, IV-4, IV-5, IV-6 and IV-7 are contained in Tables III-4 and III-5 of Appendix 3.

Table IV-2: Frequencies of actual and predicted outcomes of the ordered probit regression

Actual	Predicted	0	1	2	Total
	0	54	10	0	64
	1	13	54	12	79
	2	0	16	19	35
Total		67	80	31	178

Table IV-3: Tobit estimation results to explain sorghum farmers' variety diversification (ShaNNon)

		Marginal effects: Dy/Dx	
Variable	**Coefficient**	**Conditional on being uncensored**	**Probability uncensored**
Age	0.0017 (1.04)	0.0016	0.0004
Cht'at	-0.061^{***} (-2.99)	-0.054	-0.013
Credit*	0.02036 (0.45)	0.018	0.004
Lancrops	0.090^{***} (3.76)	0.080	0.019
Impexep	-0.058^{***} (-4.8)	-0.052	-0.012
Nopurpos	0.152^{***} (8.80)	0.135	0.032
Totlabor	0.014 (1.55)	0.012	0.003
Nonffarm*	-0.032 (-0.73)	-0.028	-0.007
Access	0.0022^{***} (2.66)	0.002	0.0005
Educate	0.0067 (0.64)	0.006	0.0014
Plots	0.0431 (1.58)	0.038	0.009
$Risk_{proxy}$	-0.00012^{**} (-2.3)	-0.0001	-0.000026
Belinarba*	0.10018 (1.32)	0.091	0.0159
Kerodeda*	0.193^{***} (2.68)	0.18	0.026
Asseliso*	0.206^{***} (2.71)	0.191	0.028
Ejeaneni*	0.302^{***} (3.53)	0.283	0.034
Chachole*	-0.070 (-0.75)	-0.06	-0.018
Gurbo*	-0.0798 (-0.76)	-0.069	-0.022
Constant	-0.245 (-2.08)		
Sigma	0.249		

Dependent variable is ShaNNon	Number of obs = 179
LR $\chi^2(18)$ = 183.15	Prob χ^2 = 0.00
Log likelihood = -5.10	Pseudo R^2 = 0.95

Notes: ***-Significant at 1%; **- Significant at 5%; and *- Significant at 10%. Values in parentheses are the ratio of the coefficient to the estimated asymptotic standard error. The method of estimation is Stata's Robust option following Huber/White standard errors and covariance. (*) Dy/Dx is for discrete change of dummy variable from 0 to 1.

Table IV-4: Tobit estimation results to explain sorghum farmers' variety diversification (SHANEVEN)

		Marginal effects: Dy/Dx	
Variable	**Coefficient**	**Conditional on being uncensored**	**Probability uncensored**
Age	0.0023 (1.27)	0.0021	0.00036
Cht'at	-0.067*** (-2.7)	-0.060	-0.0106
Credit*	-0.08 (-1.38)	-0.071	-0.0137
Lancrops	0.061*** (2.66)	0.0552	0.0096
Impexep	-0.061*** (-3.88)	-0.051	-0.0095
Nopurpos	0.155*** (7.47)	0.1394	0.0243
Totlabor	0.011 (1.02)	0.0096	0.0017
Nonffarm*	-0.053 (-0.95)	-0.048	-0.0086
Access	0.0034*** (3.34)	0.0031	0.0005
Educate	0.0041 (0.32)	0.0037	0.0006
Plots	0.030 (0.94)	0.027	0.0047
$Risk_{proxy}$	-0.000037 (-0.51)	-0.00003	-5.82e-06
Belinarba*	0.135* (1.78)	0.124	0.0153
Kerodeda*	0.114 (1.42)	0.105	0.0141
Asseliso*	0.1395 (1.49)	0.128	0.0165
Ejeaneni*	0.244*** (2.65)	0.228	0.0238
Chachole*	-0.051 (-0.48)	-0.045	-0.0091
Gurbo*	-0.149 (-1.16)	-0.127	-0.0366
Constant	-0.061 (-0.42)		
Sigma	0.298		
Dependent variable is SHANEVEN		Number of obs = 177	
LR $\chi^2(18)$ = 145.0		Prob χ^2 = 0.0	
Log likelihood = -36.9		Pseudo R^2 = 0.66	

Notes: ***-Significant at 1%; **- Significant at 5%; and *- Significant at 10%. Values in parentheses are the ratio of the coefficient to the estimated asymptotic standard error. The method of estimation is Stata's Robust option following Huber/White standard errors and covariance. (*) Dy/Dx is for discrete change of dummy variable from 0 to 1.

Table IV-5: Ordered probit regression results to explain farmers' motivations to diversify on crops (AGRODIV)

Variable	Coefficient	Marginal effects (Dy/Dx) when *Agrodiv* equals 0	1	2	3
Age	-0.01 (-1.25)	0.002	0.003	-0.003	-0.0013
Cht'at	0.28^{***} (3.1)	-0.041	-0.067	0.072	0.035
Credit*	0.198 (0.87)	-0.029	-0.048	0.052	0.0252
Lancrops	0.28^{***} (2.9)	-0.042	-0.069	0.074	0.0361
Impexep	0.045 (0.69)	-0.0067	-0.011	0.012	0.0057
Totlabor	0.10^{**} (2.16)	-0.015	-0.024	0.026	0.013
Nonffarm*	-0.22 (-1.0)	0.033	0.053	-0.058	-0.028
Access	-0.007^{*} (-1.7)	0.001	0.0017	-0.002	-0.0009
Educate	0.061 (1.30)	-0.009	-0.015	0.016	0.008
Plots	0.146 (1.30)	-0.022	-0.035	0.038	0.019
$Risk_{proxy}$	-0.0001 (-0.6)	0.00002	0.00003	-0.000035	-0.000017
Belinarba*	0.393 (1.02)	-0.047	-0.108	0.093	0.062
Kerodeda*	0.196 (0.58)	-0.027	-0.051	0.049	0.028
Asseliso*	-1.3^{***} (-3.4)	0.308	0.096	-0.308	-0.096
Ejeaneni*	-0.58 (-1.51)	0.110	0.099	-0.153	-0.056
Chachole*	-0.81^{**} (-1.9)	0.171	0.109	-0.211	-0.070
Gurbo*	-0.85 (-1.56)	0.198	0.082	-0.218	-0.062
Constant	0.560 (1.05)				

Dependent variable is AGRODIV   Number of obs. = 179
Wald (2 (17) = 104.27   Prob (2 = 0.00
Log-likelihood = -180.97   Pseudo R2 = 0.23

Notes: ***-Significant at 1%; **- Significant at 5% and *- Significant at 10%. Values in parentheses are the ratio of the coefficient to the estimated asymptotic standard error. The method of estimation is Stata's Robust option following Huber/White standard errors and covariance. (*) Dy/Dx is for discrete change of dummy variable from 0 to 1.

Table IV-6:Frequencies of actual and predicted outcomes of the ordered probit regression

Actual		Predicted 0	1	2	3	Total
	0	11	19	0	0	30
	1	4	45	20	0	69
	2	0	19	30	5	54
	3	0	4	12	10	26
Total		15	87	62	15	179

Table IV-7: Tobit regression results for proportion of land allocated to IVs: sorghum

Variable	Coefficient
Constant	-0.5212* (-1.880)
Age1	-0.0030 (-0.656)
Totlabor	-0.00065 (-0.022)
Lancrops	0.1104** (1.948)
Visits	0.42054** (2.498)
Access2	-0.6447 (-0.291)
Impexep	0.14506*** (4.888)
Nonffarm	-0.0997 (-0.873)
Educate1	-0.028 (-1.000)
$Risk_{proxy}$	0.000197 (1.088)
Credit	-0.0753 (-0.652)
Sigma	0.5963*** (9.702)
Dependent variable IMPROPOR	
Number of observations is 188	
Log likelihood -126.01	
Lower = 0 Upper = 1	

Appendix 4-B . Supplement to Chapter 6

Table IV-8: Ordered probit regression results to explain farmers' demand for variety attributes

Variable	Coeff. for yildrank	Coeff. for mrktrank	Coeff. for stabrank	Coeff. For adaprank
Howlong	-0.008*(-1.40)	-0.0005 (-0.09)	0.0003 (0.06)	-0.007 (-1.09)
Educathh	0.045** (1.62)	-0.048** (-1.94)	0.0181 (0.74)	0.0125 (0.46)
Anyincom	-0.18 (-1.03)	-0.068 (-0.40)	0.415*** (2.7)	-0.0161 (-0.09)
Accesse2	-0.0008 (-0.24)	-0.0018 (-0.47)	0.004 (1.13)	0.00172 (0.42)
Locattribute	0.28*** (5.07)	0.104** (2.09)	0.049 (0.88)	0.129** (2.46)
Cofland	-0.24 (-0.65)	0.562* (1.71)	-0.45* (-1.56)	-0.526* (-1.69)
Labindic	0.057 (0.73)	0.0299 (0.52)	-0.027 (-0.44)	0.084 (1.12)
Lanindic	-0.005 (-0.29)	0.0123 (0.61)	0.0018 (0.08)	-0.0039 (-0.24)
Regular	0.207* (1.73)	0.0842 (0.75)	0.027 (0.23)	0.2002* (1.70)
$Risk_{proxy}$	-0.0002* (-1.7)	0.00015 (1.07)	0.00007 (0.6)	0.00001 (0.08)
Capital4	0.0405 (0.82)	-0.093***(-2.68)	-0.057 (-1.17)	-0.027 (-0.66)
Govrank	-0.042 (-1.05)	0.0178 (0.43)	0.142*** (3.2)	0.0021 (0.05)
Naturank	-0.135* (-1.78)	-0.089* (-1.67)	-0.078* (-1.3)	0.2198*** (3.3)
Harro	-0.499** (-1.9)	0.402** (1.94)	0.3356 (1.21)	0.611** (2.08)
Kelaguda	-0.346 (-1.36)	0.134 (0.61)	0.1089 (0.39)	-0.038 (-0.14)
Kelokiri	-0.069 (-0.26)	0.342 (1.36)	-0.465 (-1.47)	-0.227 (-0.97)
Gibeboso	-1.383 (-3.13)	0.249 (0.64)	0.260 (0.72)	0.87*** (2.36)
Halosebe	-1.060 (-2.84)	0.032 (0.10)	0.0714 (0.22)	0.810** (2.19)
Sebekdeb	-1.194 (-3.67)	-0.265 (-0.81)	0.469 (1.45)	1.045*** (2.84)
	N = 232	N = 231	N = 226	N = 232
	Wald χ^2_{19} = 88.68	Wald χ^2_{19} = 30	Wald χ^2_{19} = 42	Wald χ^2_{19} = 146
	Prob > χ^2 = 0.0	Prob > χ^2 = 0.04	Prob > χ^2 =0.0	Prob > χ^2 = 0.0
	Pseudo R^2 = 0.11	Pseudo R^2 = 0.04	PseudoR^2=0.06	Pseudo R^2 = 0.2
	Log-LL = -261.11	Log-LL = -330.9	Log-LL = -334	Log-LL = -272.5

Notes: ***-Significant at 1%; **- Significant at 5%; and *- Significant at 10%. Two-tailed test is used for those variables the sign of the relationship is not predictable *a priori* and one-tailed test for those variables whose sign of relationship is predicted *a priori.* Figures in parentheses are the ratio of the coefficient to the estimated asymptotic standard error. The method of estimation is Stata's Robust option following Huber/White standard errors and covariance.

Table IV-9: Tobit regression results to explain farmers' effective demand for local coffee varieties

		Marginal effects: Dy/Dx	
Variable	**Coefficient**	**Conditional on being uncensored**	**Probability of uncensored**
Experenc	-0.0013 (-0.60)	-0.0007	-0.0014
Educathh	-0.013* (-1.65)	-0.0065	-0.0134
Accesse2	0.003** (2.37)	0.0015	0.0031
Lanindic	-0.008 (-1.24)	-0.004	-0.0083
Labindic	-0.004 (-0.19)	-0.0021	-0.0044
Regular	-0.36*** (-8.46)	-0.181	-0.378
$Risk_{proxy}$	-0.0001*** (-3.6)	-0.00007	-0.00014
Capita14	0.01235 (0.91)	0.0063	0.0131
Govrank	0.0245* (1.83)	0.0125	0.026
Naturank	0.0175 (0.89)	0.0089	0.0185
Beteryil	0.0706** (2.09)	0.036	0.075
Betedrot	-0.006 (-0.21)	-0.0032	-0.0068
Betestab	0.046* (1.36)	0.0236	0.0491
Betemakt	-0.135*** (-2.45)	-0.069	-0.143
Beterfod	-0.009 (-0.23)	-0.0044	-0.0091
Harro*	0.147* (1.81)	0.0687	-0.173
Kelaguda*	0.217** (2.39)	0.096	-0.267
Kelokiri*	-0.001 (-0.01)	-0.00056	0.0012
Gibeboso*	-0.267*** (-2.60)	-0.154	0.191
Halosebe*	-0.438*** (-4.70)	-0.2599	0.228
Sebekdeb*	-0.291*** (-3.29)	-0.168	0.210
Constant	1.014*** (5.52)	-0.0007	-0.0014
Dependent variable is proptres		Number of obs = 240	
LR $\chi^2(21)$ = 185.39		Prob > χ^2 = 0.0	
Pseudo R^2 = 0.56		Log likelihood = -73.9	

Notes: ***-Significant at 1%; **- Significant at 5%; and *- Significant at 10%. Two-tailed test is used for those variables the sign of the relationship is not predictable *a priori* and one-tailed test for those variables whose sign of relationship is predicted *a priori.* Figures in parentheses are the ratio of the coefficient to the estimated asymptotic standard error. (*) Dy/Dx is for discrete change of dummy variable from 0 to 1.

Appendix 4-C . Supplement to Chapter 7

Table IV-10: Treatment effects regression results to estimate the average opportunity cost of growing local sorghum varieties

Variable	Coefficient
Timarket	0.775 (0.77)
Chat1	89.3** (2.02)
Educate1	17.87 (0.76)
Allquality1	-310.91*** (-4.22)
Inputindex	-767.8*** (-3.49)
Rainfall	-191.7** (-2.02)
Plotdist	0.244 (0.13)
Visits	2.74 (0.15)
Impexep	51.71* (1.63)
Oxen	174.43*** (2.60)
Sorgindx	592.2*** (4.26)
User	659.2*** (2.96)
Cons	279.9 (0.84)
Athrho	-0.557** (-2.13)
Lnsigma	6.397*** (75.6)
Rho	-0.5058 (-0.789)
Sigma	599.9 (508.2)
Lambda	-303.46 (-561.7)
Dependent variable is Gmm	Number of obs = 166
Wald χ^2 (12) = 68.9	Prob > χ^2 = 0.00
Log likelihood = -1367.0	
Wald test of indep.equations (rho = 0): χ^2 (1) = 4.5 Prob > χ^2 = 0.033	

Notes: *** Significant at 1 percent level of significance; ** Significant at 5 percent level of significance; and *Significant at 10 percent level of significance. The method of estimation is Stata's Robust option following Huber/White standard errors and covariance.

Table IV-11: Treatment effects regression results to estimate the average opportunity cost of growing local wheat varieties

Variable	Coefficient
Laborph	0.449 (1.01)
Age	5.45 (1.53)
Sold	862.84*** (6.74)
Exteexpr	12.03 (0.26)
Inpuindx	-532.2* (-1.73)
Scoldumy	237.6*** (5.04)
Plotqulx	-63.9 (-0.77)
Rfdistri	144.3* (1.57)
Oxenph	4.38 (0.44)
Whetinde	1413.6*** (8.70)
Chemical	29.7 (0.20)
Fertph	-159.99** (-2.21)
Haresaw	-334.6 (-1.16)
Berhan	-23.76 (-0.10)
Yetmen	871.3*** (2.70)
Debrzeit	-48.22 (-0.16)
Korodega	-309.7 (-1.00)
Shasemen	59.47 (0.20)
Eteya	-83.88 (-0.28)
Durame	-280.11 (-0.79)
Improvex	687.02** (2.16)
Constant	-954.96 (-1.77)
Rho	-0.181 (-0.539)
Sigma	889.1 (827.9)
Lambda	-161.16 (-525.4)

Dependent variable is Gmboth   Number of obs = 402

Wald $\chi^2(21)$ = 275.14   Log likelihood = -3403.69

Prob > χ^2 = 0.00

LR test of indep. equations (rho = 0): χ^2 (1)= 0.77 Prob > χ^2 =0.38

Notes: *** Significant at 1 percent level of significance. ** Significant at 5 percent level of significance. *Significant at 10 percent level of significance.

Source: 1999 rural household survey by Economics Department, AAU, in collaboration with USAID

Table IV-12: Regression results of Heckman's two step sample selection model: sorghum

Variable	Coefficients (UIVs)	Coefficients (UFVs)
Constant	1455.9 (1.74)	329.97 (1.161)
Access2	-0.34 (-0.103)	-6.3*** (-4.339)
Chat1	116.0* (1.725)	75.8** (2.178)
Rainfall	-357.4*** (-3.197)	-49.9 (-0.741)
Educate1	-17.1 (-0.450)	44.8** (2.265)
Allquali	-302.6** (-2.042)	-288.6*** (-5.033)
Inputind	-870.0** (-1.908)	-522.1*** (-3.198)
Allfert	0.063 (0.458)	-0.59*** (-2.619)
Visits	-26.53 (-0.712)	
Impexep	52.15 (0.646)	
Package	13.17 (0.077)	
Oxen	221.30** (2.498)	136.87*** (2.643)
Sorgindx	709.33*** (3.494)	476.02*** (4.137)
Lambda	-415.58 (-0.867)	304.78*** (3.271)
Selection rule is: USER = 1 Selected sample from 175 is 84 Dependent variable = GMM R^2 = 0.492 Adjusted R^2 = 0.398 Model test: F[13, 70] = 5.21 Prob. value = 0.00 Log-L = -656.18 Log-L = -692.27 Correlation of disturbance in regression and selection criterion (Rho) = -0.58		Selection rule is: USER = 0 Selected sample is 91 Dependent variable = GMM R^2 = 0.58 Adjusted R^2 = 0.53 Model test: F[10, 80] = 11 Prob. value = 0.00 Log-L = -655.24 Log-L = -700.76 Correlation of disturbance in regression and selection criterion (Rho): 0.76

Notes: ***-Significant at 1%; **- Significant at 5%; and *- Significant at 10%. Values in parentheses are the ratio of the coefficient to the estimated asymptotic standard error.

Table IV-13: Regression results of Heckman's two step sample selection model: wheat

Variable	Coefficient (UIVs)	Coefficient (UFVs)
Constant	-3371.4 (-1.951)	-678.1 (-1.37)
Laborph	0.689 (1.154)	-0.340 (-0.513)
Age	24.9* (1.817)	2.9 (0.781)
Sold	101.05 (0.287)	1024.8*** (8.860)
Exteexpr	7.35 (0.081)	
Inpuindx	189.2 (0.144)	-636.9** (-2.09)
Scoldumy	232.1** (1.933)	251.6*** (5.052)
Plotqulx	261.7 (1.180)	-74 4 (-0.896)
Rfdistri	93.5 (0.373)	71.4 (0.778)
Oxenph	38.3 (1.218)	4.5 (0.455)
Whetinde	2331.9*** (2.699)	1562.1*** (9.833)
Chemical	345.4 (0.913)	-86.6 (-0.724)
Fertph	-32.5 (-0.178)	-185.4** (-2.52)
Lambda	-343.8 (-1.355)	149.4 (0.633)
Selection rule is IMPROVEX = 1 Selected sample from 409 is 50 Dependent variable = GMBOTH R^2 = 0.502 Adjusted R^2 = 0.32 Model test: F[13, 36] = 2.790 Prob. value = 0.008 Log-L = -402.54 Log-L = -428.19 Correlation of disturbance in regression and selection criterion (Rho) = -0.37		Selection rule is IMPROVEX = 1 Selected sample from 402 is 352 Dependent variable = GMBOTH R^2 = 0.36 Adjusted R^2 = 0.34 Model test: F[12, 339] = 16.23 Prob. value = 0.00 Log-L = -2881.63 Log-L = -2968.14 Correlation of disturbance in regression and selection criterion (Rho): 0.17

Notes: ***-Significant at 1%; **- Significant at 5% and *- Significant at 10%. Values in parentheses are the ratio of the coefficient to the estimated asymptotic standard error.

Source: 1999 rural household survey by Economics Department, AAU, in collaboration with USAID.

Appendix V. Shorter outline of the questionnaire used

University of Bonn, Center for Development Research
Department of Economics and Technological Change

Farm household questionnaire prepared to partly access the research project **'The economics of on-farm conservation of crop diversity in Ethiopia: incentives, attribute preferences, and opportunity costs of maintaining local varieties of crops'**[7]

May, 2001

Interview No.: _____ Date: ______ Name and sign of interviewer: ________

Part I. Basic household related information

1.0. Name of the household head ____________________________

1.1. Woreda i. Manna___ ii. Seka Chekorsa ___ iii. Goma ___1.2. PA ___ Village___

1.3. Agro-ecology of the village a. 'Dega'__ b. 'Weina dega'__ c. 'Kola'__

1.4. Sex of the household head 1. M ___2. F _____

1.5. Age of the household head ___ 1.6. Experience in farming (years) ___

1.7. Household members' roster: list of all household members

Household member[8]	**Age**	**Sex**		**Relationship with the head**[9]	**Responsibility in the household**[10]	**Level of education**[11]
		M	F			

1.8. Religion i. Muslim ____ ii. Christian ____

---

[7]Similar questionnaire was administered for sorghum. For the purpose at hand, only the one used for collecting data concerning coffee is presented.
[8]Put the household members in their age order starting with the household head.
[9]Use the following codes (put in bracket) in the table for each member of the household: Head (1), Wife or Husband (2), Son or Daughter (3), Father or Mother (4), Brother or Sister (5), Other relative (s) (6), Servant or Servant's relative (7), Tenant or Tenant's relative (8), and Other persons not related to the household head (9).
[10]Put '0' if he/she does not have any labor contribution because of age or permanent illness such as handicap.
[11]Put '0' if he/she is illiterate and 1 if he/she can read or write because of church education, quran, any informal means of getting literacy, and/or schooling upto grade 1. Put numbers for grades 2-12 and 13 for any schooling greater than high school complete.

1.9. What is the household size? ______
2.0. How important is coffee for the household? i. Most important ______ ii. Very important ____ iii. Important ____ iv. Less important ___ v. Not important ____
2.1. For what purpose(s) do you mainly grow coffee? i. Cash____ ii. Food ___ iii. Any other (please specify) _____
3.0. How do you judge the poverty status of your household in relation to other households in the area?
i. Very poor ___ ii. Poor ____ iii. Self-sufficient ____ iv. Rich ____ v. Very rich ____

Part II. Farmers' land use decisions

4.0. What is the total land size operated by the household? _____________

4.1. How many plots do you have? ________________

4.2. How much time do you need to travel on foot from one plot to the other on average? (put the average figure please) ____________________

4.3. If a neighbor has a hectare of land to be rented out, how much would you pay? (Br./Ha. or quintals of any crop per hectare) _________________

4.4. If you have a hectare of land to be rented out, how much would you accept? (Br./Ha. or quintals of any crop per hectare) _________________

5.0. Have you ever grown improved coffee variety(ies) on the farm? Yes_____ No ______

If Yes, 5.1. Which year (s)? ______ 5.2. How many are they? _____

5.3. Could you please name the variety (ies)? ________________________

If No, 5.4. Why not? ______________________________________

5.5. How much land have you planted for the improved coffee variety (ies)? ____________

5.6. From where did you get the improved variety (ies)? i. MoA ____ ii. NGO____ iii. A friend or relative ____ iv. Private seed enterprise _____ v. Any other (please specify) _______

5.7. How many improved varieties of coffee do you know so far? _______

5.8. Is your decision whether to use improved or indigenous variety (ies) influenced by the supplier? Yes_______ No ________

5.9. Do you buy (receive) improved seeds (seedlings) regularly (say every time you want to plant a new coffee tree) or was it a one time opportunity and you are re-using the seed (seedling) every time? i. Every time_____ ii. One time _____________

6.0. How many improved coffee trees do you use at the moment?_______

6.1. How many local coffee trees do you have? _______________

7.0. Coffee production and land use data by type of local variety (ies) for the last production season[12]

Name of the local or improved variety	Quantity of land used (in appropriate units)	Quality of land[13]	Quantity of production last cropping season

8.0. How much land have you planted for the improved coffee variety (ies) this cropping season? __________

8.1. How much land have you planted for the local coffee variety (ies) this cropping season? ____________

9.0. How was the rainfall distribution last cropping season? i. Good___ ii. Medium___iii. Bad ___

10.0. Quantity and types of inputs used for the improved variety (ies) of coffee and their respective price during the last cropping season? (Only those coffee trees harvested last cropping season) ___________________

11.0. Quantity and types of inputs used for the local variety (ies) of coffee and their respective price during the last cropping season? (Only those coffee trees harvested last cropping season) _____________________

Part III. On-farm conservation of indigenous coffee varieties

12.0. What are the most important production constraints that the household has been experiencing in the growing of coffee in this PA (please rank)?

i. Land _____ ii. Working capital_____ iii. Labor _____ iv. Improved seeds (seedlings) ______

v. Crop pests _______ vi. Improved farm tools ______ vii. Storage _______

12 Put the figures as single entity if the farmer is planting all varieties mixed.

13 Use the following codes: 1-good; 2-medium; and 3-bad

viii. Draught power _________ ix. Weather and drought __________ x. Extension service ______xi. Prices (of inputs and output) _________ xii. Absence of market outlet _______

xiii. Any other (please mention) ________________

13.0. What are the factors that you consider in your decision whether to use or not to use a given coffee variety (please rank)?

i. Yield _____ ii. Yield stability _____ iii. Fertilizer responsiveness ______

iv. Tolerance to environmental stress – drought, salinity, frost, wind, soil fertility _______ v. Marketability _______ vi. Disease resistance ________

vii. Food quality and taste _______ viii. Early maturity _______ ix. Other criteria (Please specify) ________

14.0. How many local varieties of coffee are grown on the plot(s) you are managing? _______

15.0. How long have you been planting coffee (in years)? i. The local variety (ies)? _______ ii. The improved variety (ies)? ________________

16.0. When have you planted coffee seed (seedling) for the last time (E.C.)? ___________

17.0. Are there any local varieties of coffee that were being used by the household but abandoned currently? Yes _________ No __________

17.1. Why did you abandon them? ________________________________

17.2. Are there any improved varieties of coffee that were being used by the household but abandoned currently? Yes _________ No __________

17.3. Why did you abandon them? ________________________________

18.0. Yield of the improved variety (ies) in the following weather (RF) scenarios

Name of the improved variety	**Yield in a very good year (Qt./Ha.)**	**Yield in an average year (Qt./Ha.)**	**Yield in a bad year (Qt./Ha.)**

18.1. Yield of the local variety (ies) in the following scenarios:

Name of the local variety	**Yield in a very good year (Qt./Ha.)**	**Yield in an average year (Qt./Ha.)**	**Yield in a bad year (Qt./Ha.)**

19.0. Price information about the local variety (ies) of coffee?

Name of the local variety	Lowest price during last cropping season (Birr / Qt.)	Highest price during the last cropping season (Birr / Qt.)

19.1. Price information about the improved variety (ies) of coffee?

Name of the improved variety	Lowest price during last cropping season (Birr / Qt.)	Highest price during the last cropping season (Birr / Qt.)

20.0. What is the minimum extra yield that you would like to get from the improved variety before you replace indigenous coffee trees by the improved ones (say for a hectare of the variety planted)? ____________

21.0. If you know in advance that the weather is bad, would you plant improved or local variety (ies)? i. Improved ___ ii. Local __ iii. Why? ____

22.0. Suppose you have two options. One is a single variety of coffee with a yield of 10 quintals per hectare and the other option is three varieties with the same productivity. Which one would you choose? i. One variety ____ ii. Three varieties _____22.1. Why ? _____

22.2. Suppose you have two options. One is a single variety of coffee with a yield potential of 12 quintals per hectare in a good season and with a yield potential of 6 quintals per hectare in a bad season. The other option is three varieties of coffee with a yield potential of 10 quintals per hectare in a good season and with a yield potential of 8 quintals per hectare in a bad season. Which one would you choose?

i. One variety _______ 2. Three varieties ______ 22.3. Why? ________

23.0. How do you rank the importance of the following variety attributes to the household?

1. a variety that gives more stable yield year after year ________
2. a variety that is more adaptable to your environment _______
3. a variety that gives very good yield during a good season _______
4. a variety that gives average yield during a bad season ________
5. a variety that fetches the highest price _________
6. a variety that is early maturing ________

23.1. If you have to make a choice among the following coffee varieties, which one would you choose?

i. A variety that is high yielding;

ii. A variety that has more stable yield;

iii. A variety that has better environmental adaptability;

iv. A variety that is marketable; and

v. A variety that is disease resistant.

24.0. At the moment, if you are given equal chance either to use the improved or the indigenous variety (ies), which one would you plant? i. Improved _______Why? ______

ii. Indigenous ______ Why? _______ iii. Both _______Why? ________

25.0. Do you want to always keep local varieties on the farm? i. Yes _____ Why? _______ ii. No _____ Why not? _______

25.1. Do you want to have as many number of varieties of coffee as possible? Yes _____ Why? ______ No ___ Why not? _____

26.0. What are your most important reasons to maintain local variety (ies) of coffee on the farm? (please rank) i. Yield _____ ii. Yield stability_____ iii. Fertilizer responsiveness _____ iv. Tolerance to environmental stress – drought, salinity, frost, wind, soil fertility _______ v. Marketability _____ vi. Disease resistance ______ vii. Food quality and taste ______ viii. Early maturity _______ ix. Other criteria (please specify) ______

27.0. What are your most important reasons to use the improved variety (ies) of coffee on the farm? (please rank) i. Yield ______ ii. Yield stability_______ iii. Fertilizer responsiveness ______

iv. Tolerance to environmental stress – drought, salinity, frost, wind, soil fertility _____v. Marketability ___vi. Disease resistance ____vii. Food quality and taste _____viii. Feed value ____ ix. Early maturity _______ x. Other criteria (please specify) _______

28.0. What are the most important factors that you consider to maintain or replace a given coffee variety on the farm? (please rank) i. Yield ______ ii. Yield stability _____ iii. Fertilizer responsiveness _______iv. Tolerance to environmental stress – drought, salinity, frost, wind, soil fertility _____

v. Marketability _______vi. Disease resistance ________ vii. Food quality and taste ________ viii. Feed value _______ ix. Early maturity _______ x. Other criteria (please specify) _______

29.0. What is the minimum quantity of output (per year) of an enterprise (Birr per year) that the household needs to survive? (Use only one most important crop or Birr per year or Birr per month)

Livestock OR crop enterprise OR Birr per year	Quantity (in the appropriate unit)	Price per unit

30.0. What do you mostly do to survive when crops fail and sustaining the life of household members becomes difficult? i. Migration to cities______ ii. Selling fixed assets _____ iii. Off-farm work _____ iv. Any other (please specify) _______

31.0. What are the annual and perennial crop enterprises that the household deals with? _______

31.1. What are the livestock enterprises that the household deals with? ___

____

32.0. Quantity of output harvested from all annual crop enterprises last cropping season

Crop enterprise	Quantity produced (in the appropriate unit)	Price per unit

32.1. Quantity of output harvested from all perennial crop enterprises except coffee last cropping season

Crop enterprise	Quantity produced (in the appropriate unit)	Price per unit

32.2. Quantity of output harvested from all livestock enterprises last production season

Livestock enterprise	Quantity produced (in the appropriate unit)	Price per unit

33.0. Quantity of land used for all annual crops last cropping season

Crop enterprise	Quantity of land used	Quality of land[14]

33.1. Quantity of land used for all perennial crops (except coffee) last cropping season

Crop enterprise	Quantity of land used	Quality of land[15]

34.0. Quantity and price of all inputs used in all perennial crops (except coffee) last cropping season

Type of input	Quantity used (in appropriate units)	Price per unit

35.0. Current livestock ownership by the household: types, quantity and price of each

Livestock	Quantity (in the appropriate unit)	Price per unit

36.0. Quantity and price of all inputs used in all livestock enterprises last cropping season

Name of the input	Quantity (in the appropriate unit)	Price per unit

Part IV. Farmers' perception towards coffee varieties and variety attributes

37.0. How do you rank the following criteria of varieties in order of importance to the household?

i. Yield ________ ii. Yield stability ___________ iii. Fertilizer responsiveness _______________

iv. Tolerance to environmental stress – (drought, salinity, frost, wind, soil fertility) _________

[14]Use the following codes: 1-good; 2-medium; and 3-bad
[15]Use the following codes: 1-good; 2-medium; and 3-bad

v. Marketability _________ vi. Disease resistance ___________

vii. Food quality and taste _____ viii. Other criteria (please specify) _____

38.0. How do you characterize the improved variety (ies) of coffee you use in terms of:

i. Yield a. Very good _____ b. Good ______ c. Poor ________

ii. Disease resistance a. Very good _____ b. Good ______ c. Poor ___

iii. Drought tolerance a. Very good _____ b. Good ______ c. Poor ______

iv. Yield stability a. Very good _____ b. Good ______ c. Poor ________

v. Marketability a. Very good _____ b. Good ______ c. Poor ________

vi. Yield without fertilizer a. Very good ____ b. Good ____ c. Poor ___

vii. Food quality and taste a. Very good ____ b. Good _____ c. Poor _____

viii. Early maturity a. Very good _____ b. Good ______ c. Poor ________

ix. Other criteria (please specify) _____ a. Very good _____ b. Good _____ c. Poor _____

38.1. How do you characterize the local variety (ies) of coffee you use in terms of:

i. Yield a. Very good _____ b. Good ______ c. Poor ________

ii. Disease resistance a. Very good _____ b. Good ______ c. Poor _______

iii. Drought tolerance a. Very good _____ b. Good ______ c. Poor _______

iv. Yield stability a. Very good _____ b. Good ______ c. Poor ________

v. Marketability a. Very good _____ b. Good ______ c. Poor ________

vi. Responsiveness to fertilizer a. Very good ____ b. Good ____ c. Poor __

vii. Food quality and taste a. Very good ____ b. Good ____ c. Poor ____

viii. Early maturity a. Very good ____ b. Good _____ c. Poor _____

ix. Other criteria (please specify) _____ a. Very good ______ b. Good ____ c. Poor ____

Part IV. Access to agricultural extension, markets and credit

39.0. Have you been participating in the extension package program? i. Yes____ ii. No____

39.1. If Yes, are you still participating? i. Yes____ ii. No____

39.2. If No, Why did you interrupt? _______________

39.3. How many times have you been visited by the extension agent last cropping season?__________

39.4. How much time do you need to travel on foot to reach the agricultural extension center? ______________

40.0. How much time do you need to travel on foot to reach the nearest dry weather road? ___________________

41.0. How much time do you need to travel on foot to reach the nearest market to buy or sell your agricultural products? _____________

41.1. What quantity of the coffee you produce last cropping season is sold? i. Local variety (ies) ___________ ii. Improved variety (ies) _______

42.0. What is the amount of coffee purchased (if any) by the household last cropping season? ______

43.0. Do you participate in any formal or informal credit scheme? i. Yes____ ii. No____

43.1. For what purpose do you want the credit? i. Purchasing inputs like fertilizer, chemicals and so on ______ ii. Purchasing improved seeds ______ iii. Purchasing household consumables _______

iv. Any other (please specify) _____________

43.2. How long have you been participating in the credit scheme(s) - in years? _____________

Part V. Household income sources

44.0. Does the household have any income source outside agriculture?

i. Yes____ ii. No_____

44.1. Do you OR any member of your household work non-farm economic activities (weavery, pottery, blacksmithing, *etc.*)? i. Yes____ ii. No_____

44.2. Do you OR any member of your household work off-farm economic activities? i. Yes___ ii. No___

45.0. What is the quantity of output you get from all off-farm and non-farm enterprises last cropping season?

Enterprise	**Quantity (in the appropriate unit)**	**Price per unit**

46.0. What is the annual cost of all inputs used in all non-farm enterprises (if any)?

Input	**Quantity (in the appropriate unit)**	**Price per unit**

47.0. Could you please rank the relative importance of the following income sources?

i. Agriculture - crop production (please specify) _________

ii. Agriculture - animal production and animal products (please specify) ____

iii. Non-farm small scale economic activities _________

iv. Off farm job opportunity ______ v. Any other (please specify) _______

VI. Optional

49.0. Any other suggestion (s); please tell us any other suggestions not mentioned above (if you have) in relation to your day to day farming activities:____

Thank you very much for sharing your practical experiences with us!

References

Adesina, A.A. and J.Baidu-Forson. 1995. Farmers' perceptions and adoption of new agricultural technology: evidence from analysis in Burkina Faso and Guinea, West Africa. *Agricultural Economics* 13: 1-9.

Adesina, A.A. and M.E. Zinnah. 1993. Technology characteristics, farmers' perceptions and adoption decisions: a tobit model application in Sierra Leone. *Agricultural Economics* 9: 297-311.

Alteiri, M.A. and C.L. Merrick. 1987. In situ conservation of crop genetic resources through maintenance of traditional farming systems. *Economic Botany* 41: 86–96.

Amemiya, T. 1984. Tobit models: a survey. *Journal of Econometrics* 24: 3-61.

Arrow, K.J. and A.C. Fisher. 1974. Environmental preservation, uncertainty, and irreversibility. *Quarterly Journal of Economics* 88: 312-319.

Baltagi, B.H. 1999. *Econometrics*. Third edition. Berlin, Germany: Springer-Verlag.

Barbier, E.B., J.C. Burgess and C. Folke. 1994. *Paradise lost? The ecological economics of biodiversity*. London, UK: Earthscan Publications Ltd.

Bellon, M.R. 1996. The dynamics of crop infraspecific diversity: a conceptual framework at the farmer level. *Economic Botany* 50 (1): 26-39.

Bellon, M.R. 1991. The ethno-ecology of maize variety management: a case study from Mexico. *Human Ecology* 19: 389-418.

Bellon, M.R., J.-L. Pham and M.T. Jackson. 1997. Genetic conservation: a role for rice farmers in plant genetic conservation. In *Plant genetic conservation: the in situ approach*. Maxted, N., B.V. Ford-Lloyd and J.G. Hawkes (Eds): 261-289. London, UK: Chapman and Hall.

Bellon, M.R. and J.E. Taylor. 1993. Farmer soil taxonomy and technology adoption. *Economic Development and Cultural Change* 41: 764-786.

Berkes, F., D. Feeny, B. McCay and J.M. Acheson. 1989. The benefits of the commons. *Nature* 340: 91-93.

Berthaud, J. and A. Charrier. 1988. Genetic resources of coffea Arabica. In *Coffee, Agronomy*, volume 4. Clarke, R.J. and R. Macrae (Eds.). London, UK: Elsevier Applied Sciences.

Bhattarai, M. and M.D. Hammig. 1998. Environmental policy analysis and instruments for biodiversity conservation: a review of recent economic literature. http//www.mtnforum.org/resources /library /bhatx98a.htm. Accessed in June 2002.

Blundell, R. and M. Consta-dias. 2000. Evaluation methods for non-experimental data. *Fiscal Studies* 21 (4): 427-468.

Blundel, R., M. Costa-Dias, C. Meghir, and J. van Reenen. 2001. Evaluating the employment impact of a mandatory job search assistance program. IFS working paper 01/20. www.ifs.org.uk. Accessed in November 2003.

Bommer, D.F.R. 1991. The historical development of international collaboration in plant genetic resources. In *Searching for new concepts for collaborative genetic resources management*. van Hintum, T.J.L., L. Frese and P.M. Perret (Eds.): 3-12. Rome, Italy: International Board for Plant Genetic Resource.

von Braun, J. and D. Virchow. 1996. Economic Evaluation of Biotechnology and Biodiversity in Developing Countries. *Agriculture and Rural Development* 3, 1.

Bromley, D.W. 1992. Entitlements and public policy in environmental risks. In *The social response to environmental risk: policy formulation in an age of uncertainty*. Bromley, D.W. and K. Segerson (Eds.): 1-21. Boston, USA: Kluwer Academic Publishers.

Brookfield, H. and M. Stocking. 1999. Agrodiversity: definition, description and design. *Global Environmental Change* 9: 77-80.

Brown, W.L. 1983. Genetic diversity and genetic vulnerability – an appraisal. *Economic Botany* 37 (1): 4-12.

Brown, W.L. 1988. Plant genetic resources: a view from the seed industry. In *Seeds and sovereignty: the use and control of plant genetic resources*. Kloppenburg, J.R. (Ed.): 218-230. Durham, USA: Duke University Press.

Brush, S.B. 2000. The issues of in situ conservation of CGRs. In *Genes in the field: on-farm conservation of crop diversity*. Brush, S.B. (Ed.): 3-26. Ann Arbor, USA: Lewis Publishers.

Brush, S.B. 1995. In situ conservation of landraces in centers of crop diversity. *Crop Science* 35: 346-354.

Brush, S.B. 1992a. Farmers' rights and genetic conservation in traditional farming systems. *World Development* 20 (11): 1617-1630.

Brush, S.B. 1992b. Reconsidering the green revolution: diversity and stability in cradle areas of crop domestication. *Human Ecology* 20 (2): 145-167.

Brush, S.B. 1991. A farmer-based approach to conserving crop germplasm. *Economic Botany* 45: 153-165.

Brush, S.B. 1986. Genetic diversity and conservation in traditional farming systems. *Journal of Ethnobiology* 6: 151-67.

Brush, S.B. and E. Meng. 1998. The value of wheat genetic resources to farmers in Turkey. In *Agricultural values of plant genetic resources*. Everson, R.E., D. Gollin and V. Santaniello (Eds.): 97-113. Wallingford, UK: CABI Publishing.

Brush, S.B., J.E. Taylor and M. R. Bellon. 1992. Technology adoption and biological diversity in Andean potato agriculture. *Journal of Development Economics* 39: 365-387.

Brush, S.B., M.B. Corrales, and E. Schmidt. 1988. Agricultural development and maize diversity in Mexico. *Human Ecology* 16 (3): 307-328.

Burstin, J., M. Lefort, M. Mittaeu, A. Sontot, and J. Guiard. 1997. Towards the assessment of the cost of gene banks management: conservation, regeneration, and characterization. *Plant Varieties and Seeds* 10: 163-172.

Cameron, A.C. and P.K. Trivedi. 1998. *Regression analysis of count data*. Cambridge, UK: Cambridge University Press.

Cameron, A.C. and P.K. Trivedi. 1990. Regression-based tests for overdisperssion in the Poisson model. *Journal of Econometrics* 46: 347-364.

Cameron, A. C. and P.K. Trivedi. 1986. Econometric models based on count data: comparisons and applications of some estimators and tests. *Journal of Applied Econometrics* 1: 29-54.

Chochie field gene bank site manager. 2002. Interview held in November, 2002.

Conelly, W.T. and M.S. Chaiken. 2000. Intensive farming, agro-diversity, and food security under conditions of extreme population pressure in Western Kenya. *Human Ecology* 28 (1): 19-51.

Cornes, R. and T. Sandler. 1986. *The theory of externalities, public goods and club goods*. First edition. Cambridge, UK: Cambridge University Press.

Correa, C.M. 2000. In situ conservation and intellectual property rights. In *Genes in the field: on-farm conservation of crop diversity*. Brush, S.B. (Ed.): 239-260. Ann Arbor, USA: Lewis publishers.

CRDA. 2002. *Food security, sustainable agriculture and trade*. Addis Ababa, Ethiopia: Christian Relief and Development Association.

Cromwell, E. and S. van Oosterhout. 2000. On-farm conservation of crop diversity: policy and institutional lessons from Zimbabwe. In *Genes in the field: on-farm conservation of crop diversity*. Brush, S.B. (Ed.): 217-238. Ann Arbor, USA: Lewis publishers.

CTA. 1999. *Ethiopia: cradle of the wonder bean - Coffee Arabica*. Addis Ababa, Ethiopia: Coffee and Tea Authority.

CSA. *Statistical Bulletins, various issues*. Addis Ababa, Ethiopia: Central Statistical Authority.

Dasgupta, P. 1996. The economics of the environment. *Environment and Development Economics* 1: 387-421.

Dasgupta, P. 1995. The population problem: theory and evidence. *Journal of Economic Literature* 33: 1897-1902.

Dasgupta, P. 1993. *An inquiry into wellbeing and destitution*. Oxford, UK: Oxford University Press.

Debela, S. 1988. Agricultural research and food production in Ethiopia. *Ethiopian Journal of Agricultural Sciences* 10 (1-2): 25-37.

Dehejia, R.H. and S. Wahba. 1998. *Propensity score matching methods for non-experimental causal studies*. NBER working paper No. 6829.

Demissie, A. 2001. Biodiversity conservation of medicinal plants: problems and prospects. In *Conservation and sustainable use of medicinal plants in Ethiopia*. Zewdu, M. and A. Demissie (Eds.): 56-64. Addis Ababa, Ethiopia: Institute of Biodiversity Conservation and Research.

Demissie, A. 1991. A decade of germplasm exploration and collecting activities by the Plant Genetic Resources Center/Ethiopia. In *Plant genetic resources of Ethiopia*. Engels, K.M.M., J.G. Hawkes and W. Melaku (Eds.): 202-217. Cambridge, UK: Cambridge University Press.

Demissie, A. and T. Arega. 2000. *A dynamic farmer-based approach to the conservation of Ethiopia's plant genetic resources*. Progress report 1997/1998. Addis Ababa, Ethiopia: Institute of Biodiversity Conservation and Research.

Dercon, S. 1996. Risk, crop choice, and savings: evidence from Tanzania. *Economic Development and Cultural Change* 45: 485-513.

Doggett, H. 1991. Sorghum history in relation to Ethiopia. In *Plant Genetic Resources of Ethiopia*. Engels, J. M. M., J. G. Hawkes, and M. Worede (Eds.): 140-159. Cambridge, UK: Cambridge University Press.

Dubale, P. and D. Teketay. 1999. *The need for forest coffee germplasm conservation in Ethiopia and its significance in the control of coffee diseases*. Paper presented at the Coffee Berry Disease Workshop, 13-15 August, 1999. Addis Ababa, Ethiopia: Ethiopian Agricultural Research Organisation.

Dulloo, M.E., L. Guarino, F. Engelmann, N. Maxted, H.J. Newbury, F. Attere, and B.V. Ford Lloyd. 1999. Complementary conservation strategies for the genus coffea with special reference to the Mascarene Islands. *Genetic Resources and Crop Evolution* 45: 565-579.

van Dusen, M.E. 2000. *In situ conservation of crop genetic resources in the Mexican Milpa System*. Ph.D. dissertation, Berkeley, USA: University of California.

Edilegnaw, W.Z. 2003. The opportunity cost of growing traditional wheat varieties: implications for the design of targeting principles for adoption of improved varieties in Ethiopia. In *Proceedings of the national workshop on technological progress in Ethiopian Agriculture*. Demeke, M., Alemu M., Assefa A. and Dejene A. (Eds.): 63-86. Addis Ababa, Ethiopia: Addis Ababa Univertsity.

Edilegnaw, W.Z., J. Mburu, and D. Virchow. 2003. Incentives, opportunity costs and contract design for on-farm conservation in Ethiopia. Paper presented at the third BioEcon workshop, May 22 – 25, 2003, France, Montepellier. http://www.bioecon.ucl.ac.uk/bioecon.htm. Accessed in September, 2003.

Ehrenfeld, D.W. 1976. The conservation of non-resources. *American Scientist* 64:648–56.

Ehrenfeld, D.W. 1988. Why put a value on biodiversity? In *Biodiversity*. Wilson, E.O. (Ed.): 212-216. Washington, D.C., USA: National Academic Press.

Ellis, R.H., T.D. Hong, E.H. Roberts and K.-L. Tao. 1990. Low moisture content limits to relations between seed longevity and moisture. *Annals of Botany* 65: 493-504.

Emerton, L. 2000. Using incentives for biodiversity conservation. The World Conservation Union http://biodiversityeconomics.org/incentives. Accessed in May 2000.

Engels, J.M.M. and J.G. Hawkes. 1991. The Ethiopian gene centre and its genetic diversity. In *Plant genetic resources of Ethiopia*. Engels, K.M.M., J.G. Hawkes and W. Melaku (Eds.): 23-41. Cambridge, UK: Cambridge University Press.

Engels, J.M.M. and D. Wood. 1999. Conservation of agrobiodiversity. In *Agrobiodiversity: characterization, utilization and management.* Wood, D. and J.M. Lenne (Eds.): 355-385. Wallingford, UK: CABI Publishing.

Epperson, J.E., D.H. Pachico and C.L. Guevara. 1997. A cost analysis of maintaining cassava plant genetic resources. *Crop Science* 37: 1641-1649.

Espinosa, J.A. and B.K. Goodwin. 1991. Hedonic price estimation for Kansas wheat characteristics. *Western Journal of Agricultural Economics* 16 (1): 72-85.

Evenson, R.E., D. Gollin and V. Santaniello. 1998. Introduction and overview: agricultural values of plant genetic resources. In *Agricultural values of plant genetic resources*. Evenson, R.E., D. Gollin and V. Santaniello (Eds.): 1-12. Wallingford, UK: CABI Publishing.

FAO. 1998. *The state of the world's genetic resources for food and agriculture*. Rome, Italy: FAO.

FAO. 1989. *Plant genetic resources*. Rome, Italy: FAO.

FDRE. 1998. *National Policy on Biodiversity Conservation and Research*. Addis Ababa, Ethiopia: Environmental Protection Authority.

FDRE. 1997. *Environmental Policy*. Addis Ababa, Ethiopia: Environmental Protection Authority in collaboration with Ministry of Economic Development and Cooperation.

FDRE. 1996. Country report to the International Technical Conference on Plant Genetic Resources, 17-23 June 1996, Leipzig, Germany.

Feder, G., R.E. Just and D. Zilberman. 1985. Adoption of agricultural innovations in developing countries: a survey. *Economic Development and Cultural Change* 33: 255-298.

Feder, G. and D.L. Umali. 1993. The adoption of agricultural innovations: a review. *Technological Forecasting and Social Change* 43: 215-239.

Feyissa, R. 2002. Experiences with community seed banks in Ethiopia. In *Incentive measures for sustainable use and conservation of agrobiodiversity: experiences and lessons from Southern Africa*. Almekinders, C.J.M. (Ed.): 35-40. Wageningen, The Netherlands: Technical Centre for Agricultural and Rural Cooperation.

Fox, W. 1984. *The intuition to deep ecology*. The ecologist.

Freeman, H. A., S.K Ehui and M.A. Jabbar. 1998. Credit constraint and smallholder dairy production in the East African highlands: application of a switching regression model. *Agricultural Economics* 19:33-34.

Frisvold, G.B. and P.T. Condon. 1998. The convention on biological diversity and agriculture: implications and unresolved debates. *World Development* 26 (4): 551-570.

Furubotn, E.G. and S. Pejovich. 1974. Introduction. In *The economics of property rights*. Furubotn, E.G. and S. Pejovich (Eds.): 1-9. Cambridge, UK: Ballinger Publishing.

Gari, J.A. 1999. *Biodiversity conservation and use: local and global considerations*. MA, USA: Harvard University Press.

Gasson, R. 1973. Goals and values of farmers. *Journal of Agricultural Economics* 24: 521-542.

Gebrekidan, B. 1979. Sorghum genetic resources in Africa. *Ethiopian Journal of Agricultural Sciences* 1 (2): 108-125.

Gebrekidan, B. and Y. Kebede. 1979. The traditional culture and yield potentials of the Ethiopian high lysine sorghums. *Ethiopian Journal of Agricultural Sciences* 1 (1): 29-40.

Gebre-Mariam, H. 1991. Wheat production and research in Ethiopia. In *Wheat research in Ethiopia: a historical perspective*. Gebre-Mariam, H., D.G. Tanner and M. Hulluka (Eds.): 1-15. Addis Ababa, Ethiopia: IAR/CIMMYT.

Goldfeld, S. and R. Quandt. 1973. The switching of structural shifts by switching regressions. *Annals of Economic and Social Measurement* 2 (4): 475-485.

Gole, T.W. 2002. Conservation and use of coffee genetic resources in Ethiopia: challenges and opportunities in the context of current global situations. http.//www.gdnet.org/pdf. Accessed in June 2003.

Gole, T.W., M. Denich, T. Demel, and P.LG. Vlek. 2002. Human impacts on coffea arabica genetic diversity in Ethiopia and the need for its in situ conservation. In *Managing plant genetic diversity: proceedings of the international conference on science and technology for managing plant genetic diversity in the 21st century*. J. Engels, V.R. Rao, A.H.D. Brown and M. Jackson (Eds.): 237-247. Rome, Italy: IPGRI.

Gollin, D. M. Smale, and B. Skovmand. 1998. Optimal collection and search for crop genetic resources. In *Farmers, gene banks and crop breeding: economic analyses of diversity in wheat, maize, and rice*. Smale, M. (Ed.): 57-78. CIMMYT and Kluwer Academic publishers.

Gourieroux, C., A. Monfort and A. Trognon. 1984. Pseudo maximum likelihood methods: applications to poisson models. *Econometrica* 52 (3): 701-720.

Greene, W. 2000. *Econometric Analysis*. Fourth Edition. New Jersey, USA: Prentice-Hall.

Hardon, J. 1996. National Sovereignty and access to genetic resources. *Biotechnology and Development Monitor* 27: 24.

Hardon, J. and W.S. de Boef. 1993. Linking farmers and breeders in local development. In *Cultivating knowledge*. de Boef, W., K. Amanor, K. Wellard and A. Bebbington (Eds.): 1-13. London, UK: Intermediate Technology Publications.

Harlan, J.R. 1969. Ethiopia: a center of diversity. *Economic Botany* 23 (4): 309-314.

Hartell, J., M. Smale, P.W. Heisey, and B. Senauer. 1998. The contribution of genetic resources and diversity to wheat productivity in the Punjab of Pakistan. In *Farmers, gene banks and crop breeding: economic analyses of diversity in wheat, maize, and rice*. Smale, M. (Ed.): 145-158. CIMMYT and Kluwer Academic Publishers.

Hawkes, J. G., N. Maxted and B. V. Ford-Lloyd. 2000. *The ex situ conservation of plant genetic resources*. New York, USA: Kluwer Academic Publishers.

Hawtin, G. C. and T. Hodgkin. 1997. Towards the future. In *Plant genetic conservation: the in situ approach*. Maxted, N., B.V. Ford-Lloyd and J.G. Hawkes (Eds.): 368-383. London, UK: Chapman and Hall.

Hayami, Y. and V. Ruttan. 1985. *Agricultural development: an international perspective*. Baltimore, USA: Johns Hopkins University Press.

Hazell, P. 2002. Why invest in less-favored areas. www.ruralforum.info/papers/hazell1En.pdf. Accessed in May 2003.

Heckman, J. 1997. Instrumental variables: a study of implicit behavioural assumptions used in making program evaluation. *Journal of Human Resources* 32: 441-462.

Heckman, J. 1979. Sample selection as a specification error. *Econometrica* 47 (1): 153-161.

Heckman, J. and J. Hotz. 1989. Choosing among alternative non-experimental methods for estimating the impact of social programs: the case of manpower training. *Journal of the American Statistical Association* 84 (408): 862-880.

Heckman, J., H. Ichimura and P. Todd. 1998. Matching as an econometric evaluation estimator. *Review of Economic Studies* 65 (2): 261-294.

Heckman, J., H. Ichimura and P. Todd. 1997. Matching as an econometric evaluation estimator: evidence from evaluating a job training program. *Review of Economic studies* 65 (4): 605-654.

Heckman, J. and R. Robb. 1985. Alternative methods for evaluating the impact of interventions. In *Longitudnal analysis of labor market data.* Heckman, J. and B. Singer (Eds.): 156-244. New York, USA: John Wiley.

Heckman, J., J. Smith and N. Clements. 1997. Making the most out of programme evaluations and social experiments: accounting for heterogeneity in programme impacts. *Review of Economic Studies* 64 (4): 487-535.

Heisey, P.W., M. Smale, D. Byerlee and E. Souza. 1997. Wheat rusts and the costs of genetic diversity in the Punjab of Pakistan. *American Journal of Agricultural Economics* 79: 726-737.

Hellerstein, D. and R. Mendelsohn. 1993. A theoretical foundation for count data models. *American Journal of Agricultural Economics* 75: 604-611.

Hendler, R. 1975. Lancaster's new approach to consumer demand and its limitations. *The Economic Journal* 65 (1): 194-199.

Heywood, V.H. (Ed.). 1995. *Global biodiversity assessment*. Cambridge, USA: Cambridge University Press.

Holling, C.S. 1973. Resilience and stability of ecological systems. *Annual Review of Ecology and Systematics* 4: 1-23.

Huang, C.L., R. Raunikar and S. Misra. 1991. The application and economic interpretation of selectivity models. *American Journal of Agricultural Economics* 73 (2): 496-502.

IBCR. 2003. Official documentation.

IBPGR. 1991. *Elsevier's Dictionary of Plant Genetic Resources*. Amsterdam, The Netherlands: Elsevier Science Publishers.

IDRC. 1995. Ethiopia's living laboratory of biodiversity. http://www.IDRC.ca/books/mag/12ethiop.html. Accessed in April 2003.

IUCN. 1999. Building Economics into national biodiversity strategies and action plan – experiences from Eastern Africa. Paper prepared for workshop on building biodiversity into sectoral strategies and action plans, 18-20 June 1999, Montreal, Canada.

IUCN. 1980. *World conservation strategy: living resource conservation for sustainable development*. Gland, Switzerland: IUCN-UNEP-WWF.

Jain, H.K. 1994. The biodiversity convention: more losers than winners. *Biotechnology and Development Monitor* 21: 24.

de Janvry, A., M. Fafchamps, and E. Sadoulet. 1991. Peasant household behavior with missing markets: some paradoxes explained. *The Economic Journal* 10: 1400-1417.

Jarvis, D. and T. Hodgkin. 2000. Farmer decision making and genetic diversity: linking multidisciplinary research to implementation on-farm. In *Genes in the field: on-farm conservation of crop diversity*. Brush, S.B. (Ed.): 261-278. Ann Arbor, USA: Lewis publishers.

Johnston, J. and J. DiNardo. 1997. *Econometric methods*. Fourth Edition. New York, USA: The McGraw-Hill.

Just, R.E. and D. Zilberman. 1983. Stochastic structure, farm size, and technology adoption in developing agriculture. *Oxford Economic Papers* 35: 307-328.

Karper, R. E. and J. R. Quinby. 1947. Sorghum - its production, utilization, and breeding. *Economic Botany* 1: 355-371.

Kebede, Y. 1991. The role of Ethiopian sorghum germplasm resources in the national breeding programme. In *Plant genetic resources of Ethiopia*. Engels, K.M.M., J.G. Hawkes and W. Melaku (Eds.): 315-322. Cambridge, UK: Cambridge University Press.

Kidane, A. and D.G. Abler. 1994. Production technologies in Ethiopian agriculture. *Agricultural Economics* 10: 179-191.

Koo, B. and B.D. Wright. 1999. The role of biodiversity products as incentives for conserving biological diversity: some instructive examples. *The Science of the Total Environment* 240: 21-30.

Krutila, J.V. 1967. Conservation reconsidered. *The American Economic Review* 57: 777-786.

Ladd, G.W. and V. Suvannunt. 1976. A model of consumer goods characteristics. *American Journal of Agricultural Economics* 58: 504-510.

Laffont, J-J. and J. Tirole. 1993. *A theory of incentives in procurement and regulation*. Massachesetts, USA: Massachesetts Institute of Technology.

Lancaster, K.J. 1966. A new approach to consumer theory. *Journal of Political Economy* 74: 132-57.

Lipton, M. and R. Longhurst. 1989. *New seeds and poor people*. Baltimore, USA: The John Hopkins University Press.

Maddala, G.S. 1983. *Limited dependent and qualitative variables in econometrics*. Cambridge, UK: Cambridge University Press.

Magurran, A. 1988. *Ecological diversity and its measurement*. New Jersey, USA: Princeton University Press.

Maier, L and M. Shobayashi. 2001. *Multifunctionality: towards an analytical framework*. Paris, France: Organization for Economic Co-operation and Development.

Mann, C.C. 1991. Extinction: are ecologists crying wolf? *Science* 253: 736-738.

Maxted, N., J.G. Hawkes, B.V. Ford-Lloyd and J.T. Williams. 1997. A practical model for in situ genetic conservation. In *Plant genetic conservation: the in situ approach*. N. Maxted, B.V. Ford-Lloyd and J.G. Hawkes (Eds.): 339-367. London, UK: Chapman and Hall.

McDonald, J.F. and R.A. Moffitt. 1980. The uses of Tobit analysis. *Review of Economics and Statistics* 62: 318-321.

McFadden, D.L. 1974. The measurement of urban travel demand. *Journal of Public Economics* 3: 303-328.

McNeely, J. A. 1988. *Economics and biological diversity: developing and using economic incentives to conserve biological resources*. Gland, Switzerland: International Union for Conservation of Nature and Natural Resources.

McNeely, J.A. and F. Vorhies. 2000. Economics and conserving forest genetic diversity. http://www.biodiversityeconomics.org/pdf/topics-05-01.pdf. Accessed July 2003.

Meng, E.C.H. 1997. *Land allocation and in situ conservation of crop genetic resources: the case of wheat landraces in Turkey*. Ph.D. dissertation, Davis, USA: University of California.

Meng, E.C.H., J.E. Taylor and S.B. Brush. 1998. Implications for the conservation of wheat landraces in Turkey from a household model of varietal choice. In *Farmers, gene banks and crop breeding: economic analyses of diversity in wheat, maize, and rice*. Smale, M. (Ed.): 127-142. CIMMYT and Kluwer Academic Publishers.

Mesfin, A. 2002. Personal communication, October 2002.

Ministry of coffee and tea development. 1984. Coffee Statistics handbook, 1960/61-1982/83. Addis Ababa, Ethiopia: Government Printing Office.

Morris, M.L. and P.W. Heisey. 1998. Achieving desirable levels of crop diversity in farmers' fields: factors affecting the production and use of commercial seed. In *Farmers, gene banks and crop breeding: economic analyses of diversity in wheat, maize, and rice*. Smale, M. (Ed.): 217-238. CIMMYT and Kluwer Academic Publishers.

Mountford, H. and J.H. Keppler. 1999. Financing incentives for the protection of biodiversity. *The Science of the Total Environment* 240: 133-144.

Mugabe, J. and E. Ouko. 1994. Control over genetic resources. *Biotechnology and Development Monitor* 21: 6-7.

Mugwara, R.T. and T.B. Gebre Egtiabher. 2002. Incentive measures to enhance the sustainable use of agrobiodiversity. In *Incentive measures for sustainable use and conservation of agrobiodiversity: experiences and lessons from Southern Africa.* Almekinders, C.J.M. (Ed.). Proceeding of a workshop, Lusaka, Zambia, 11-14 September, 2001.

Mulualem, T. 1999. *Morphological and molecular diversity in durum wheat. Triticum durum Desf. landraces of North Shewa*. MSc thesis, Addis Ababa, Ethiopia: Addis Ababa University.

Murduch, J. 1995. Income smoothing and consumption smoothing. *Journal of Economic Perspectives* 9 (3): 103-114.

Negatu, W. and A. Parikh. 1999. The impact of perception and other factors on the adoption of agricultural technology in the Moret and Jiru Woreda. district of Ethiopia. *Agricultural Economics* 21: 205-216.

Nehring, K. and C. Puppe. 2002. A theory of diversity. *Econometrica* 70 (3): 1155-1198.

Netting, R.M. and Stone, M.P. 1996. Agro-diversity on a farming frontier: Kofyar smallholders on the benue plains of Central Nigeria. *Africa* 66 (1): 52-70.

Norgaard, R.B. 1985. Traditional agricultural knowledge: past performance, future prospects, and institutional implications. *American Journal of Agricultural Economics* 66: 874-8.

Norton-Griffiths, M. and C. Southey. 1995. The opportunity cost of biodiversity conservation in Kenya. *Ecological Economics* 12 (2): 125-139.

North, D.C. 1990. *Institutions, institutional change and economic performance*. Cambridge, UK: Cambridge University Press.

NRS (1989). *Recommended dietary allowances*. National research service, sub-committee of the 10th editions of the RDAs. Washington D.C.: National Academy Press.

NSIA. 2001. *Crop variety register*. Issue No.4. Addis Ababa, Ethiopia: National Seed Industry Agency.

OECD. 1999. *Handbook of incentive measures for biodiversity: design and implementation*. Paris, France: Organisation for Economic Cooperation and Development.

OECD. 1998. *The economics of benefit sharing: introduction and practical experiences*. Paris, France: Organisation for Economic Cooperation and Development.

OECD. 1997. *Issues in the sharing of benefits arising out of the utilization of genetic resources*. Paris, France: Organisation for Economic Cooperation and Development.

Oldfield, M.L. and J.B. Alcorn. 1987. Conservation of traditional agroecosystems: can age-old farming practices effectively conserve crop genetic resources? *Bioscience* 37: 199-208.

Omiti, J.M., K.A. Parton, S.K. Ehui and J.A. Sinden. 2000. Some policy implications of the resurfacing of rural factor markets following agrarian de-collectivization in Ethiopia. *Human Ecology* 4: 585-603.

Orr, D.W. 1992. Pascal's Wager and economics in a hotter time. *Ecological Economics* 6 (1): 1-6.

Pannel, D.J., B.Malcolm, and R.S.Kingwell. 2000. Are we risking too much? Perspectives on risk in farm modeling. *Agricultural economics* 23: 69-78.

Pardey, P.G., B. Koo, B.D. Wright, M. E. van Dusen, B. Skovmand, and S. Taba. 1999. Costing the conservation of genetic resources: CIMMYT's ex situ maize and wheat collection. *Crop Science* 41: 1286-1299.

Parikh, A. and A. Bernard. 1988. Impact of risk on HYV adoption in Bangladesh. *Agricultural Economics* 2: 167-178.

Pearce, D. and D. Moran. 1994. *The economic value of biodiversity*. London, UK: Earthscan Publications Ltd.

Perales, H.R., S.B. Brush, and C.O. Qualset. 1998. Agronomic and economic competitiveness of maize landraces and in situ conservation in Mexico. In *Farmers, gene banks and crop breeding: economic analyses of diversity in wheat, maize, and rice.* Smale, M. (Ed.): 109-127. CIMMYT and Kluwer Academic publishers.

Perlman, D. L. and G. Adelson. 1997. *Biodiversity: exploring values and priorities in conservation.* Oxford, UK: Blackwell Science.

Perrings, C. 1995. Biodiversity conservation as insurance. In *The economics and ecology of biodiversity decline: the forces driving global change.* Swanson, T.M. (Ed.): 69-77. Cambridge, UK: Cambridge University Press: Cambridge.

Perrings, C. and H. Opschoor. 1994. The loss of biological diversity: some policy implications. *Environmental and Resource Economics* 4 (1): 1-11.

Perrings, C., K-G. Mäler, C. Folke, C.S. Holling, and B-O. Jansson. 1995. Unanswered questions. In *Biodiversity loss: economic and ecological issues*. Perrings, C., K-G. Mäler, C. Folke, C.S. Holling, and B. Jansson (Eds.). Cambridge, UK: Cambridge University Press.

Pindyck, R. 1991. Irreversibility, uncertainty, and investment. *Journal of Economic Literature* 29: 1110-1148.

Pindyck and Rubinfeld. 1998. *Econometric models and economic forecasts.* Fourth Edition. Blacklick, USA: Irwin McGraw-Hill.

Pingali, P.L., S.D. Rozelle and R.V. Gerpacio. 2001. The farmer's voice in priority setting: a cross-country experiment in eliciting technological preferences. *Economic Development and Cultural Change* 49 (3): 591-609.

Prakash, T.N. and D. Virchow. 2003. Effects of incentive systems on on-farm management of crop genetic resources and required contract designs: the case of rice in India. Paper presented at the third BioEcon workshop, May 22 – 25, 2003, Montepellier, France. http://www.bioecon.ucl.ac.uk/bioecon.htm. Accessed in July 2003.

Pudney, S. 1989. *Modelling individual choice: the econometrics of corners, kinks and holes.* Oxford, UK: Basil Blackwell.

Puppe, C. 2003. Personal communication, September 2003.

Quandt, R.E. 1988. *The econometrics of disequilibrium*. Oxford, UK: Basil Blackwell.

Ratchford, B.T. 1975. The new economic theory of consumer behavior: an interpretive essay. *Journal of Consumer Research* 2: 65-75.

Rice, R. A. 2001. Noble goals and challenging terrain: organic and fair trade coffee movements in the global market place. *Journal of Agricultural and Environmental Ethics* 14: 39-66.

Rosenbaum, P.R. and D. B. Rubin. 1983. The central role of the propensity score in observational studies for causal effects. *Biometrika* 70 (1): 41-55.

Rosenzweig, M.R. and H.P. Binswanger. 1993. Wealth, wealth risk and the composition and profitability of agricultural investments. *The Economic Journal* 103: 56-78.

Rossi, P.H., H.E. Freeman and M. W. Lipsey. 1999. *Evaluation: a systematic approach*. Sixth Edition. Newbury Park, USA: Sage Publications.

Roy, A.D. 1952. Safety first and the holding of assets. *Econometrica* 20: 431- 449.

Sedjo, R. A. 1992. Property rights, genetic resources, and biotechnological change. *Journal of Law and Economics* 35: 199-213.

Sedjo, R.A. and R.D. Simpson. 1995. Property rights, externalities and biodiversity. In *The economics and ecology of biodiversity decline: the forces driving global change*. Swanson, T.M. (Ed.): 79-88. Cambridge, UK: Cambridge University Press.

Segerson, K. 1992. The policy response to risk and risk perceptions. In *The social response to environmental risk: policy formulation in an age of uncertainty*. Bromley, D.W. and K. Segerson (Eds.): 101-130. Boston, USA: Kluwer Academic Publishers.

Shands, H.L. 1991. Complementarity of in situ and ex situ germplasm conservation from the stand point of the future user. *Israel Journal of Botany* 40: 521-528.

Simpson, R.D., R.A. Sedjo and J.W. Reid. 1996. Valuing biodiversity for use in pharmaceutical research. *Journal of Political Economy* 104 (1): 163-185.

Singh, I., J. Subramanian. 1986. Agricultural household modeling in a multicrop environment: case studies in Korea and Nigeria. In *Agricultural household models: extensions, applications and policy.*

Singh, I., L. Squire, and J. Strauss (Eds.): 95-115. USA, Baltimore: John Hopkins University Press.

Singh, I., L. Squire and J. Strauss 1986 (Eds.). 1986. *Agricultural household models: extensions, applications and policy.* Baltimore, USA: John Hopkins University Press.

Smale, M. 2003. Economic incentives for conserving crop genetic diversity on farmers' fields: issues and evidence. In *Efficient conservation of crop genetic diversity: theoretical approaches and empirical studies.* Virchow, D. (Ed.): 55-68. Berlin, Germany: Springer-Verlag.

Smale, M. 1997. The green revolution and wheat genetic diversity: some unfounded assumptions. *World Development* 25 (8): 1257-1269.

Smale, M., M. Bellon, and P.L. Pingali. 1998. Farmers, gene banks and crop breeding: introduction and overview. In *Farmers, gene banks and crop breeding: economic analyses of diversity in wheat, maize, and rice.* Smale, M. (Ed.): 3-18. CIMMYT and Kluwer Academic Publishers.

Smale, M., D. Soleri, D.A. Cleveland, D. Louette, E.B. Rice and A. Aguirre. 1998. Collaborative plant breeding as an incentive for on-farm conservation of genetic resources: economic issues from studies in mexico. In *Farmers, gene banks and crop breeding: economic analyses of diversity in wheat, maize, and rice.* Smale, M. (Ed.): 239-258. CIMMYT and Kluwer Academic Publishers.

Smale, M., R.E. Just and H.D. Leathers. 1994. Land allocation in HYV adoption models: an investigation of alternative explanations. *American Journal of Agricultural Economics* 76: 535-546.

Smale, M., M. Bellon, and J.A.A. Gömez. 2001. Maize diversity, variety attributes and farmers' choices in Southeastern Guanajuato, Mexico. *Economic Development and Cultural Change* 50 (1): 201-225.

Smale, M. and M.R. Bellon. 1999. A conceptual framework for valuing on-farm genetic resources. In *Agrobiodiversity: characterization, utilization and management.* Wood, D. and J.M. Lenne (Eds.): 387-408. Wallingford, UK: CABI Publishing.

Smith, C.E. Jr. 1969. From Vavilov to the pesent – a review. *Economic Botany* 23: 2-19.

Solow, R. 1974. The economics of resources and the resources of economics. *American Economic Review* 64: 1-12.

Solow, A., S. Polasky and J. Broadus. 1993. On the measurement of biological diversity. *Journal of Environmental Economics and Management* 24: 60-68.

StatCorp. 2001. Stata 7: Users' guide.

Swaminathan, M.S. (Ed.). 1996. *Agro-biodiversity and farmers' rights*. New Delhi, India: Vedams eBooks.

Swanson, T. 1999. Conserving global biological diversity by encouraging alternative development paths: can development coexist with diversity? *Biodiversity and conservation* 8: 29-44.

Swanson, T.M. 1995. Uniformity in development and the decline of biological diversity. In *The economics and ecology of biodiversity decline: the forces driving global change.* Swanson, T.M., (Ed.): 41-53. Cambridge, UK: Cambridge University Press.

Swanson, T. and T. Göschl. 2000a. Optimal genetic resource conservation: in situ and ex situ. In *Genes in the field: on-farm conservation of crop diversity*. Brush, S.B. (Ed.): 165-191. Ann Arbor, USA: Lewis Publishers.

Swanson, T.M. and T. Göschl. 2000b. Property rights issues involving plant genetic resources: implications of ownership for economic efficiency. *Ecological Economics* 32: 75-92.

Sylvain, P.G. 1958. Ethiopian coffee – its significance to world coffee problems. *Economic Botany* 12: 111-139.

Taillie, C. and G.P. Patil. 1982. Diversity as a concept and its measurement. *Journal of the American Statistical Association* 77 (379): 548-561.

Tefera, N. and M. Demeke. 2003. The productivity and profitability of wheat and teff technologies: comparison between extension and non-extension plots in three selected villages of Ethiopia. In *Proceedings of the national workshop on technological progress in Ethiopian Agriculture*. Demeke, M., Alemu M., Assefa A. and D. Aredo (Eds.): 135-172. Addis Ababa, Ethiopia: Addis Ababa Univertsity.

Teshome, A., L. Fahrig, J.K. Torrance, J.D.H. Lambert, J.T. Arnason, and B. Baum. 1999. Maintenance of Sorghum. Sorghum Bicolor, Poaceae landrace diversity by farmers' selection in Ethiopia. *Economic Botany* 53 (1): 69-78.

Tewolde, B.G.E. 1990. The importance of Ethiopian forests in the conservation of Arabica coffee gene-pool. *Mitt. inst. allg. bot.* 23a: 65-72.

Thies, E. 2000. *Inventive measures appropriate to enhance the conservation and sustainable use of agro-biodiversity*. Eschborn, Germany: GTZ.

Thurstone, L. 1927. A law of comparative judgement. *Psychological Review* 34: 273-286.

Tisdell, C. 1990. Economics and the debate about preservation of species, crop varieties and genetic diversity. *Ecological Economics* 2: 77-90.

Train, K.E. 2003. *Discrete choice methods with simulation*. Cambridge, UK: Cambridge University Press.

UNEP. 2000. *Sustainable life on earth: how the convention on biological diversity promotes nature and human well-being*. Montreal, Canada: Secretariat of the Convention on Biological Diversity.

UNEP. 1994. *Convention on Biological Diversity*. Montreal, Canada: Secretariat of the Convention on Biological Diversity.

UNEP. 1992. *Agenda 21: Environment and Development Agenda*. United Nations Conference on Environment and Development, Rio de Janeiro, 3-14 June.

Vaughan, D.A. and Chang, T.T. 1992. In situ conservation of rice genetic resources. *Economic Botany* 46: 368-383.

Vella, F. 1998. Estimating models with sample selection bias: a survey. *The Journal of Human Resources* 32: 127-169.

Verbeek, M. 2000. *A guide to modern econometrics*. West Sussex, UK: John Wiley and Sons.

Virchow, D. (Ed.). 2003. *Efficient conservation of crop genetic diversity: theoretical approaches and empirical studies*. Berlin, Germany: Springer-Verlag.

Virchow, D. 1999a. *Conservation of genetic resources: costs and implications for a sustainable utilization of plant genetic resources for food and agriculture*. Berlin, Germany: Springer-Verlag.

Virchow, D. 1999b. Spending on conservation of plant genetic resources for food and agriculture: how much and how efficient? ZEF-Discussion Papers on Development Policy 16/1999. Bonn, Germany: University of Bonn, Center for Development Research.

Weitzman, M.L. 1993. What to preserve? An application of diversity theory to crane conservation. *Quarterly Journal of Economics* 108: 1557-1583.

Weitzman, M.L. 1992. On diversity. *Quarterly Journal of Economics* 107 (2): 363-405.

Wells, M.P. 1998. Institutions and incentives for biodiversity conservation. *Biodiversity and Conservation* 7: 815-835.

de Wet, J. M. J. and J. R. Harlan. 1971. The origin and domestication of sorghum bicolor. *Economic Botany* 25: 128-135.

White, T.A. and C.F. Runge. 1995. The emergence and evolution of collective action: lessons from watershed management in Haiti. *World Development* 23 (10): 1683-1689.

Winkelmann, R. 1995. Duration dependence and dispersion in count data models. *Journal of Business and Economic Statistics* 13: 467-474.

Witt, S.C. 1985. *Brief book: biotechnology and genetic diversity*. San Francisco, USA: California Agricultural Lands Project.

Wood, D. and J.M. Lenne. 1997. The conservation of agrobiodiversity on-farm: questioning the emerging paradigm. *Biodiversity and Conservation* 6: 109-129.

Wood, A., P. Stedman-Edwards, and J. Mang. 2000. *The root causes of biodiversity loss*. London, UK: Earthscan Publications Ltd.

Worede, M. 1997. Ethiopian in situ conservation. In *Plant genetic conservation: the in situ approach*. Maxted, N., B.V. Ford-Lloyd and J.G. Hawkes (Eds.): 290-301. London, UK: Chapman and Hall.

Worede, M. 1991. An Ethiopian perspective on conservation and utilization of plant genetic resources. In *Plant genetic resources of Ethiopia*. Engels, K.M.M., J.G. Hawkes and W. Melaku (Eds.): 1-19. Cambridge, UK: Cambridge University Press.

Worede, M. 1988. Diversity and the genetic resource base. *Ethiopian Journal of Agricultural Sciences* 10 (1-2): 39-52.

Worede, M. 1985. *Crop genetic resource activities in Ethiopia*. Paper presented at the international symposium on South East Asian plant genetic resources, August 20-24, Indonesia, Jakarta.

Worede, M. and H. Getu. undated. *Proposal for the establishment of a coffee genetic resources conservation program in Ethiopia*. Mimeographed.

Worede, M., T. Tesemma, and R. Feyissa. 2000. Keeping diversity alive: an Ethiopian perspective. In *Genes in the field: on-farm conservation of crop diversity*. Brush, S.B. (Ed.): 143-161. Ann Arbor, USA: Lewis Publishers.

Yibrah, H-S. 1998. The role of biotechnology in conservation and utilization of biological diversity. In *Biotechnology for conservation and utilization of biological resources*. Yibrah, H-S. and A. Demissie (Eds.): 89-110. Addis Ababa, Ethiopia: Institute of Biodiversity Conservation and Research.

Yibrah, H-S. and A. Demissie. 1998. Eds. Biotechnology for conservation and utilization of biological resources. Addis Ababa, Ethiopia: Institute of Biodiversity Conservation and Research.

Zavoina, T. and W. McElvey. 1975. A statistical model for analysis of ordinal level dependent variables. *Journal of Mathematical Sociology* 4: 103-20.

Zeller, M., C. Lapenu, B. Minten, E. Ralison, D. Randrianaivo, and C. Randrianarisoa. 1999. Pathways of rural development in Madagascar: an empirical investigation of the critical triangle between environmental sustainability, economic growth and poverty alleviation. *Quarterly Journal of International Agriculture* 2: 105-127.

Zeller, M., A. Diagne, and C. Mataya. 1998. Market access by smallholder farmers in Malawi: implications for technology adoption, agricultural productivity and crop income. *Agricultural Economics* 19: 219-229.

Development Economics and Policy

Series edited by Franz Heidhues and Joachim von Braun

Band 1 Andrea Fadani: Agricultural Price Policy and Export and Food Production in Cameroon. A Farming Systems Analysis of Pricing Policies. The Case of Coffee-Based Farming Systems. 1999.

Band 2 Heike Michelsen: Auswirkungen der Währungsunion auf den Strukturanpassungsprozeß der Länder der afrikanischen Franc-Zone. 1995.

Band 3 Stephan Bea: Direktinvestitionen in Entwicklungsländern. Auswirkungen von Stabilisierungsmaßnahmen und Strukturreformen in Mexiko. 1995.

Band 4 Franz Heidhues / François Kamajou: Agricultural Policy Analysis – Proceedings of an International Seminar, held at the University of Dschang, Cameroon on May 26 and 27 1994, funded by the European Union under the Science and Technology Program (STD). 1996.

Band 5 Elke M. Förster: Protection or Liberalization? A Policy Analysis of the Korean Beef Sector. 1996.

Band 6 Gertrud Schrieder: The Role of Rural Finance for Food Security of the Poor in Cameroon. 1996.

Band 7 Nestor R. Ahoyo Adjovi: Economie des Systèmes de Production intégrant la Culture de Riz au Sud du Bénin: Potentialités, Contraintes et Perspectives. 1996.

Band 8 Jenny Müller: Income Distribution in the Agricultural Sector of Thailand. Empirical Analysis and Policy Options. 1996.

Band 9 Michael Brüntrup: Agricultural Price Policy and its Impact on Production, Income, Employment and the Adoption of Innovations. A Farming Systems Based Analysis of Cotton Policy in Northern Benin. 1997.

Band 10 Justin Bomda: Déterminants de l'Epargne et du Crédit, et leurs Implications pour le Développement du Système Financier Rural au Cameroun. 1998.

Band 11 John M. Msuya: Nutrition Improvement Projects in Tanzania: Implementation, Determinants of Performance, and Policy Implications. 1998.

Band 12 Andreas Neef: Auswirkungen von Bodenrechtswandel auf Ressourcennutzung und wirtschaftliches Verhalten von Kleinbauern in Niger und Benin. 1999.

Band 13 Susanna Wolf (ed.): The Future of EU-ACP Relations. 1999.

Band 14 Franz Heidhues / Gertrud Schrieder (eds.): Romania – Rural Finance in Transition Economies. 2000.

Band 15 Katinka Weinberger: Women's Participation. An Economic Analysis in Rural Chad and Pakistan. 2000.

Band 16 Christof Batzlen: Migration and Economic Development. Remittances and Investments in South Asia: A Case Study of Pakistan. 2000.

Band 17 Matin Qaim: Potential Impacts of Crop Biotechnology in Developing Countries. 2000.

Band 18 Jean Senahoun: Programmes d'ajustement structurel, sécurité alimentaire et durabilité agricole. Une approche d'analyse intégrée, appliquée au Bénin. 2001.

Band 19 Torsten Feldbrügge: Economics of Emergency Relief Management in Developing Countries. With Case Studies on Food Relief in Angola and Mozambique. 2001.

Band 20 Claudia Ringler: Optimal Allocation and Use of Water Resources in the Mekong River Basin: Multi-Country and Intersectoral Analyses. 2001.

Band 21 Arnim Kuhn: Handelskosten und regionale (Des-)Integration. Russlands Agrarmärkte in der Transformation. 2001.

Band 22 Ortrun Anne Gronski: Stock Markets and Economic Growth. Evidence from South Africa. 2001.

Band 23 Patrick Webb / Katinka Weinberger (eds.): Women Farmers. Enhancing Rights, Recognition and Productivity. 2001.

Band 24 Mingzhi Sheng: Lebensmittelkonsum und -konsumtrends in China. Eine empirische Analyse auf der Basis ökonometrischer Nachfragemodelle. 2002.

Band 25 Maria Iskandarani: Economics of Household Water Security in Jordan. 2002.

Band 26 Romeo Bertolini: Telecommunication Services in Sub-Saharan Africa. An Analysis of Access and Use in the Southern Volta Region in Ghana. 2002.

Band 27 Dietrich Müller-Falcke: Use and Impact of Information and Communication Technologies in Developing Countries' Small Businesses. Evidence from Indian Small Scale Industry. 2002.

Band 28 Wolfram Erhardt: Financial Markets for Small Enterprises in Urban and Rural Northern Thailand. Empirical Analysis on the Demand for and Supply of Financial Services, with Particular Emphasis on the Determinants of Credit Access and Borrower Transaction Costs. 2002.

Band 29 Wensheng Wang: The Impact of Information and Communication Technologies on Farm Households in China. 2002.

Band 30 Shyamal K. Chowdhury: Institutional and Welfare Aspects of the Provision and Use of Information and Communication Technologies in the Rural Areas of Bangladesh and Peru. 2002.

Band 31 Annette Luibrand: Transition in Vietnam. Impact of the Rural Reform Process on an Ethnic Minority. 2002.

Band 32 Felix Ankomah Asante: Economic Analysis of Decentralisation in Rural Ghana. 2003.

Band 33 Chodechai Suwanaporn: Determinants of Bank Lending in Thailand: An Empirical Examination for the Years 1992 to 1996. 2003.

Band 34 Abay Asfaw: Costs of Illness, Demand for Medical Care, and the Prospect of Community Health Insurance Schemes in the Rural Areas of Ethiopia. 2003.

Band 35 Gi-Soon Song: The Impact of Information and Communication Technologies (ICTs) on Rural Households. A Holistic Approach Applied to the Case of Lao People's Democratic Republic. 2003.

Band 36 Daniela Lohlein: An Economic Analysis of Public Good Provision in Rural Russia. The Case of Education and Health Care. 2003.

Band 37 Johannes Woelcke. Bio-Economics of Sustainable Land Management in Uganda. 2003.

Band 38 Susanne M. Ziemek: The Economics of Volunteer Labor Supply. An Application to Countries of a Different Development Level. 2003.

Band 39 Doris Wiesmann: An International Nutrition Index. Concept and Analyses of Food Insecurity and Undernutrition at Country Levels. 2004.

Band 40 Isaac Osei-Akoto: The Economics of Rural Health Insurance. The Effects of Formal and Informal Risk-Sharing Shemes in Ghana. 2004.

Band 41 Yuansheng Jiang: Health Insurance Demand and Health Risk Management in Rural China. 2004.

www.peterlang.de